P9-EDQ-015

ASSAULT ON THE ROCKIES

ENVIRONMENTAL CONTROVERSIES IN ALBERTA

Canadian Cataloguing in Publication Data

Main entry under title:
Assault on the Rockies

ISBN 1-895836-49-2

1. Environment degradation – Rocky Mountains, Canadian (B.C. and Alta)*
I. Urquhart, Ian T. (Ian Thomas), 1955-
GE160.C3A87 1998 333.78'137'09711 C98-910711-6

Editor for the press: Larry Pratt
Book design by Diane Jensen
Printed and bound in Canada by Nisku Printers (1980) Ltd.

Rowan Books gratefully acknowledges the support of the Canada Council for the Arts for our publishing programme. We also acknowledge the support of the City of Edmonton and the Edmonton Arts Council.

Published in Canada by Rowan Books
an imprint of The Books Collective
214-21 10405 Jasper Avenue
Edmonton, Alberta T5J 3S2
(403) 421-1544

ASSAULT ON THE ROCKIES

ENVIRONMENTAL CONTROVERSIES IN ALBERTA

EDITED BY IAN URQUHART

EDMONTON

Printed on Resolve Premium Opaque. This paper is EcoLogo certified. The Ecologo is awarded to those papers manufacturers who meet or exceed stringent requirements for recycled fibre content as set by the Canadian Government's Environmental Choice Program. Resolve Premium Opaque contains a minimum of 50% recycled paper by weight of the finished product including 10% post-consumer fibre. The product also meets current E.P.A. recommended minimums for recycled fibre content. This paper has been manufactured using an alkaline paper making process. It meets the American National Standards Institute's (ANSI) requirements for alkaline permanence. (Alkaline is an acid free process).

For my daughters Andrea and Kali

The following photos illustrate how dramatically – and quickly – the landscapes of Alberta's Foothills have been altered by industrial use over the past half-century. The outlined black wedge shape provides a common reference point for these photos, taken between 1949 and 1991. The photos are of the Foothills north of Whitecourt. The 1949 photo shows the area as wilderness. By 1964, the development of the Judy Creek oilfield was well underway (note that the white/grey lines are roads, pipelines, or transmission lines; the small white squares are wellsites). In 1982, logging of the area had begun (the large white/grey patches are clearcuts). Finally, in 1991 logging activities had intensified – the darkest grey areas in the photo were the remaining fragments of the forested wilderness photographed in 1949.

THE VANISHING FOOTHILLS WILDERNESS

AS 120 #125

Air Photo Services, Government of Alberta

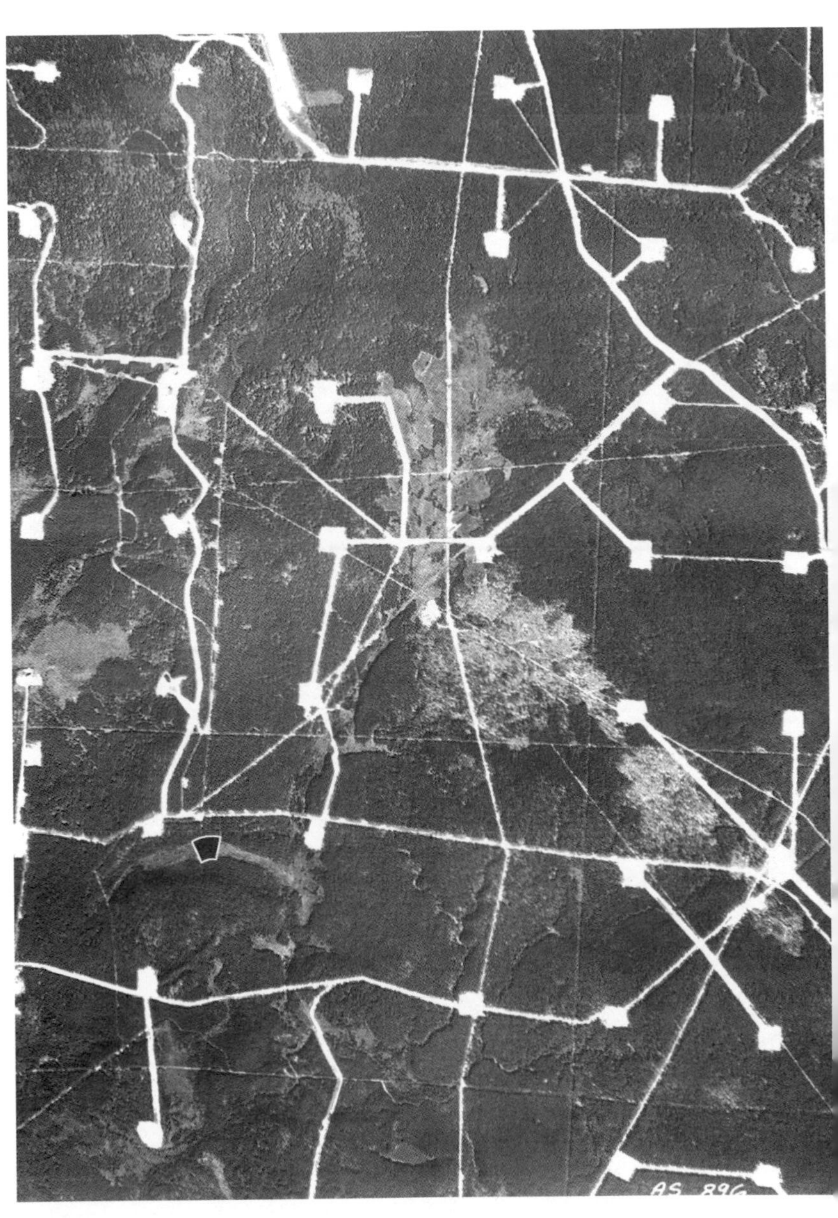

AS 896 #110

Air Photo Services, Government of Alberta

AS2527 #13

Air Photo Services, Government of Alberta

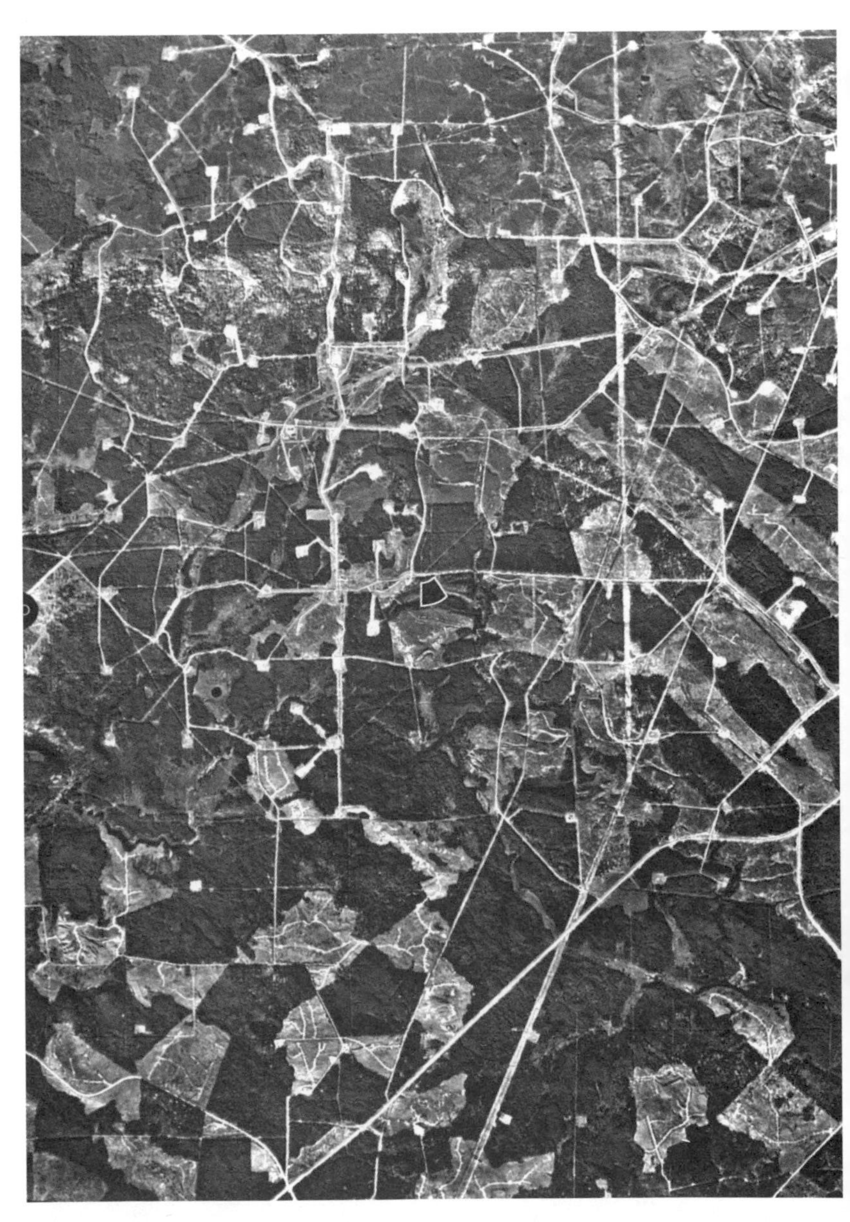

AS4209 #115

Air Photo Services, Government of Alberta

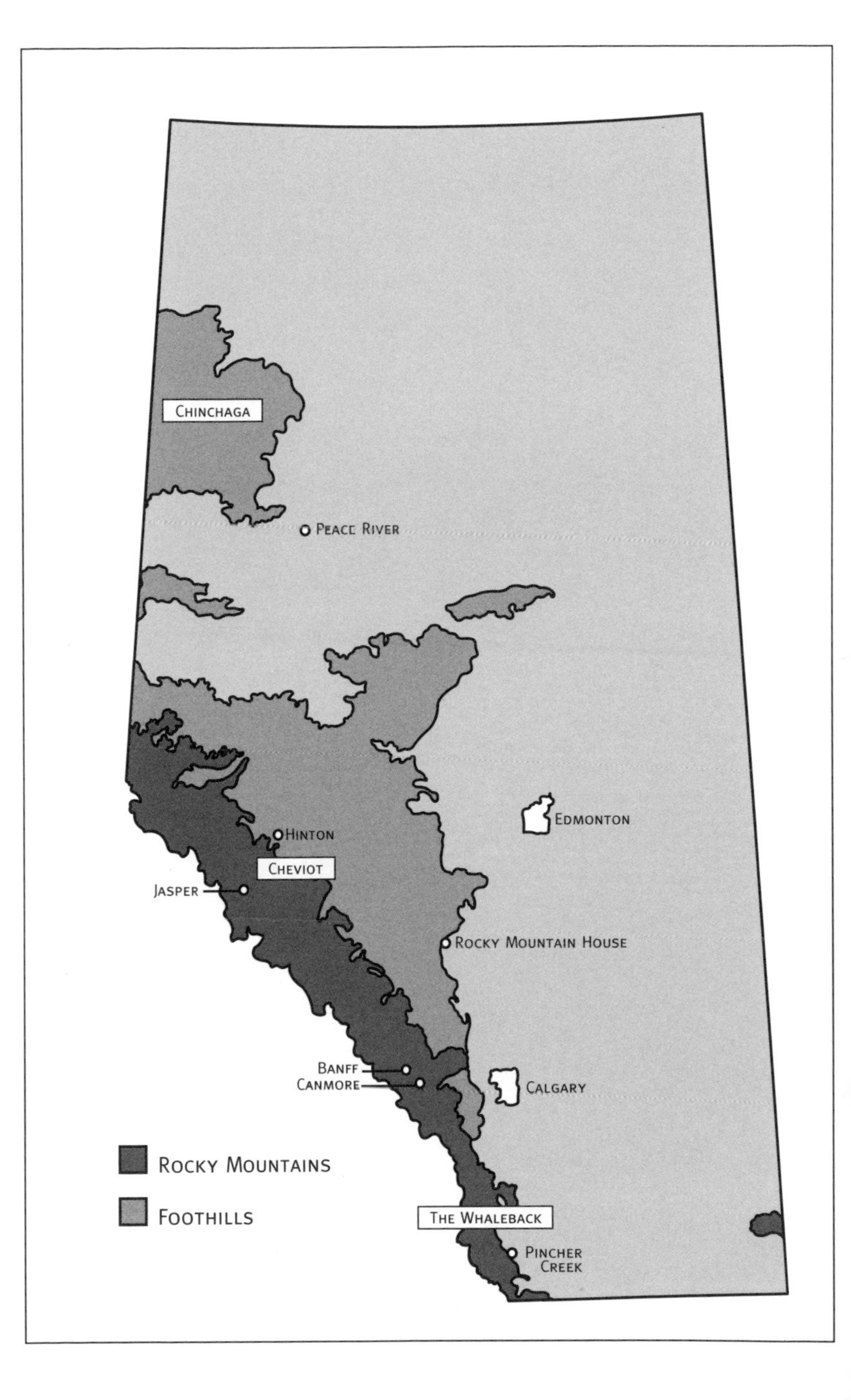
Chinchaga
Peace River
Edmonton
Hinton
Cheviot
Jasper
Rocky Mountain House
Banff
Canmore
Calgary
Rocky Mountains
Foothills
The Whaleback
Pincher Creek

TABLE OF CONTENTS

ACKNOWLEDGEMENTS

This project could not have proceeded without the support of many people and organizations. First, I would like to thank Shirley Serviss and Larry Pratt of Rowan Books for their belief in the importance of this collection. Larry's counsel and encouragement over the life of this project were very much appreciated. Second, the financial assistance I received from the Alberta Foundation for the Arts and the Faculty of Arts at the University of Alberta was very welcome and I thank them for that support. Thanks to Miriam Koene for the research help she offered during the final stages of the project. Most importantly, I would like to offer my deepest thanks to the writers, activists, publishers, and commentators whose essays and comments comprise most of this collection. This book never would have materialized without their generous participation. Finally, I would like to thank Theresa for her love, patience, and generosity.

PREFACE

Assault on the Rockies is an anthology of debates, viewpoints and prognoses on the pressures for growth and the counter-pressures for environmental protection in Alberta's foothills and in the Rocky mountains themselves. University of Alberta political scientist Ian Urquhart has done a superb job of collecting, organizing and editing a broad range of documents – poems, essays, articles, editorials and so on – written from different standpoints on what might be called the pressure points of contention in the foothills and mountains.

Urquhart is from the 'show, don't tell' school of persuasion. By that I mean that he is less interested in preaching hot gospel than in making the documentary evidence available, allowing people to make up their minds about what should happen in, say, the Whaleback or in Banff and Jasper. These issues are complex and there will be no cheap solutions. Some of the pressures for growth and development in the foothills have been there for half a century or more – I mean the oil and gas explorers, the coal companies, the forestry industry, all of which have plenty of economic and political clout and an interest in using the foothills' immense natural wealth for corporate growth. Natural gas is a big part of the story. I can recall a time when much of the Eastern Slope was regarded as off-limits to extensive corporate development, but that was before most of the major petroleum companies left and were replaced by Calgary's gas-hungry producers eager to export more natural gas to the United States. Look at the remarkable growth and profitability of the new rising stars of the oilpatch from the late Eighties to 1997 – companies like Canadian 88, Renaissance, Poco and others - and you will have part of the reason for the growth pressures on the foothills.

But it is not just the big resource corporations that want to exploit the foothills, and it is not only the tourist industry and the developers who wish to 'develop' Banff and other mountain resorts. There is also the relentless pressure of people, especially the urban middle classes in Alberta's growing cities, who want to buy up land in a beautiful natural setting and develop new suburbs in the foothills. It is the old story: with the arrival of the city dweller, rural property becomes so costly that those who have actually been working the land are forced to leave it. Understandably, many urban people have a longing to escape to the simpler, peaceful life of the country – to "arise and go now, and go to Innisfree,/And a small cabin build there..../And live alone in the bee-loud glade"[1] – but they won't find it in the foothills.The parvenu new money of urban Alberta seem hell-bent on turning the spectacular western reaches of the province into their own playground, and they too have political clout.

In contrast, the environmental movement in Alberta has had an uphill struggle in Canada's most conservative province, and the environmentalists themselves have not always been willing to turn out in sufficient numbers or to raise enough hell to prevent the big battalions of growth from having their way. When they can actually scare a few politicians, environmentalists will begin to slow the 'assault on the Rockies'. But will it be too late?

I want to thank those who helped Rowan Books with this anthology – Ian Urquhart, above all, for his meticulous and time-consuming work as editor. Ian's commitment to good writing and interesting books matches his passion for the rivers, streams and out-of-the-way lonely places of Alberta and BC. We are very grateful to those who gave permission for us to use their writings. Allan Markin, Calgary businessman and friend of the literary arts, helped us with a financial contribution, for which we thank him (would that there were more like him). And we gratefully acknowledge the support of the Canada Council and the Alberta Foundation for the Arts. Diane Jensen designed both the cover and the text with her usual skill, and my colleague Shirley Serviss has been generous in her support for, and belief in, this important book.

Larry Pratt

CHAPTER 1

INTRODUCTION

The Canadian Art Tree

which in a commercial silk screen print
    hangs on memory's wall of my Grade V classroom
    and which you remember as searingly
    as you remember
    intensely childhood injustice,
"The West Wind" or "The Jack Pine"
    depicts particularity of treeness,
    the anxiety and stress of growing up
    bent by wind while seeking moisture
    through roots clawed into Shield-hard granite
    and creating shadow where moss might hold it longer.

Every time a painter
    returns to look at a tree again
    to find a way to paint it,
    or photograph it in true and honest weather,
    the tree becomes an x-ray version
    of our forsaken blasted psyche
    which Thompson, (sic) Jackson, Carr, and my Uncle Pete
    tried to warn us against,

while in more civilized climates
    artists and parents are warning
    their children about mad people, confidence men,
    pornographers,
    we're trying to grow up straight
    where wind, snow, soil, moss and land conspire
    to shape us interesting.

by Jon Whyte[1]

To grow up straight or to be shaped interesting by climate and landscape – the imagery in Jon Whyte's poem speaks well to the tensions at the heart of this book, tensions between economy and environment. These tensions are a key feature of life in Alberta at the end of the Twentieth Century. In 1996, the federal government's Banff-Bow Valley Task Force selected the image of the crossroads to describe Banff, Canada's oldest and most famous national park. Tourism, the park's economic engine, was threatening to run the park's ecological integrity off of the tracks. Such collisions are taking place with increasing frequency and intensity throughout the Rocky Mountain and Foothills regions of Alberta. Miners, oilmen, loggers, ranchers, and a new generation of settlers all want to bend the mountains and the foothills to their purposes.

In the shadow of Jasper National Park, coal miners want to expand their open-pit mines to strip millions of tons of coal from the mountain landscape. Far to the south, where the prairies run into the mountains near the American border, grizzlies struggle to survive in an environment which is also crucial to the health of generations of ranchers. A little further north, oil and gas companies hope to give their rigs a taste of the undeveloped parcels of Rocky Mountain montane landscape found in the Whaleback. In the central and northern foothills, timber companies look to convert nature's greenery into greenbacks. Finally, there are those of us who want to escape the city, who want to move to a foothills acreage or own a cottage in the mountains, and who, if we make the move, stress the environments and traditional lifestyles of these areas. This collection focuses upon these assaults on the Rockies and the foothills.

Collectively, the Rockies and the Foothills Natural Regions cover more than twenty-two per cent of Alberta – nearly 150,000 square kilometres. Many might believe that, despite the pressures noted above, this impressive amount of territory is well-protected and, to a considerable degree, pristine. They would be mistaken. Our century long search for economic growth in Alberta has left very few corners of these ecosystems untouched by development. The Foothills, Alberta's third largest natural region, is also our third most altered region. A few statistics make the point. By the end of

1995, at least 203 natural gas fields and 28 oil fields had been discovered lying, in whole or in part, beneath the Foothills. To unlock these petroleum riches, 26,906 wellsites existed.[2] As Table 1 underlines dramatically, the activities of the energy industry and forestry, a second key economic sector, have left Alberta's Foothills without virtually any wilderness at all – 0.39 per cent!

While the magnitude of this loss of wilderness may not be evident to all, Albertans care about preserving some of the wilderness which is a crucial part of the province's heritage. In 1994, the World Wildlife Fund (WWF) polled Albertans about whether or not they wanted to see portions of landscapes like the Rockies and the Foothills protected from all types of industrial activities. Ninety-three per cent favoured total protection for a proportion of these landscapes. Of course, as in many surveys of this type, they were not asked if they favoured total protection if it came at the expense of jobs – especially their own. Nonetheless, the figure is a strong endorsement of the campaign of the World Wildlife Fund and the Canadian Parks and Wilderness Society (CPAWS) to secure total protection for a small percentage of these ecosystems.

Several of the more significant controversies featured in this collection – the Chinchaga, the Whaleback – have been targeted for protection by groups like the WWF, CPAWS, and the Alberta Wilderness Association (AWA). Their controversial status is due, in part, to the way the provincial government of Alberta has formulated a protected areas strategy. This strategy – Special Places 2000 – is described by the government as a "made-in-Alberta" strategy to establish a network of representative landscapes across the province. Its "made-in-Alberta" flavour may come from its fidelity to Alberta's history of supporting natural resource development. Special Places does not believe in total protection. Instead, it wants "to balance the goal of preservation, with the parallel goals of outdoor recreation, heritage appreciation, and tourism/economic development." In Alberta, an area could be protected and host all of these activities. As Table 2, the WWF's report card on Alberta's progress towards securing a network of protected areas suggests, the WWF does not think much of Alberta's efforts. As Ty Lund's comments on Alberta's 1997-98 grade emphasize, Alberta's Minister of Environmental Protection does not think much of the WWF.

TABLE 1: How Much Wilderness is Left in Alberta's Foothills?

	NUMBER	PERCENTAGE OF FOOTHILLS TOWNSHIPS
Total number of Foothills Townships	1,297	100
Townships without logging on Crown Land(up to 1994/95)	573	44.18
Townships without wellsites	114	8.79
Townships without Crown Land logging and wellsites	91	7.02
Townships without Crown Land logging, wellsites, and linear disturbances*	5	0.39

What is wilderness? According to the Concise Oxford Dictionary of Ecology, wilderness is "an extensive area of land which has never been permanently occupied by humans or subjected to their intensive use (e.g., for mineral extraction or cultivation) and which exists in a natural or nearly natural state." Alberta's Department of Environmental Protection developed a practical test of how much wilderness still exists in the Foothills. They identified all Foothills townships affected by the energy or the forest products industry (a township equals thirty-six square miles or 93.23 square kilometres).

Of the 94,790.23 square kilometres covered by the Foothills, only five townships – approximately 400 square kilometres – had not been affected by the energy or forestry industries. As the department said, it used this simple approach "to graphically demonstrate the extent to which human activity has permeated and affected" Alberta's Foothills.

* linear disturbances refer to roads, pipeline/transmission line right of ways, or more than five kilometres of seismic cutlines.
Source: Alberta Environmental Protection.

TABLE 2: Alberta's Endangered Spaces Report Cards

1997-98: F
1996-97: D+
1995-96: B
1994-95: F
1993-94: C
1992-93: D

Ty Lund: "I don't know what they're talking about. We can't figure how on earth they come to these grades. I mean, let's get with the program. This is so ridiculous. It's an insult to the province... This special interest group, centred in Toronto, comes and says the sky is falling."

In selecting or writing material for this collection I tried to ensure that the reader will encounter a variety of perspectives on the particular issue or practice discussed. Discussions on Banff or Cheviot or grizzlies offer opposing points of view. Criticisms of a logging road approval in the foothills are followed by defences of clearcutting – the controversial logging practice companies use there. This effort to allow you to hear "both sides" of an issue does not mean that I do not have my biases. I do. But, I believe that the purpose of the collection is best promoted by ensuring that the views you disagree with also are presented.

Several other features of the collection merit explanation. At first glance, some of the older selections may strike you as "dated." I do not believe that this conclusion will stand up to a close reading of these selections. Older material is included if it is thought-provoking, provides important background material to contemporary disputes, and describes a reality we still live. The material reprinted here passes this test easily.

The structure of the collection, its use of poetry at the beginning of most chapters, was chosen because the natural settings we consider here are emotional landscapes as well. As the fur trader Alexander Henry is claimed to have said on a February morning in 1811 when he looked up the North Saskatchewan River towards the Rockies:

> At eleven o'clock we came to Swallow Rock, the spot whence our Columbia canoes turned back on the 16th of last October. From this spot a long reach up river opens to the view, with a grand sight of the mountains ahead, covered with snow. Here the green pines on both sides of the river seem to end, and all above appears to be one continuous dreary waste, destroyed by fire some years ago. The ground is covered with immense piles of fallen trees lying across one another, which gives a gloomy appearance to the country, while towering summits of the mountains strike the mind with awe.[3]

The Rockies and the Foothills may inspire us, they may depress us, but, they will always impress us.

Finally, the editor's contributions to this collection, other than postscripts, are initialled "I.U."

C H A P T E R 2

PRESSURES FOR GROWTH

Storm

reminds us, makes us mind our minds,
slickens to glistening our fears
because of its glittering darkness
from the west
    makes mountains out of sky
manifests itself in anger
making mythic the wrath we thought personal
until terror, no more discrete,
in wind rising, raising a roar of waterfall
till all we hoped still
becomes judgement
opaque in tremulous clouds
and, we sense, mind will not mind
bu sky's ire and passion
drumming the skin of mountains
arms pounding
    the march of
storm, batter it into beauty

- by Jon Whyte[1]

"But a conservation effort that concentrates only on the extremes of industrial abuse tends to suggest that the only abuses are the extreme ones when, in fact, the earth is probably suffering more from many small abuses than from a few large ones."

– Wendell Berry[2]

— INTRODUCTION —

Through two cases, the suburbanization of the foothills and energy exploration north of Waterton National Park, we meet the types of pressures facing Alberta's Rockies and Foothills today. Joshua Avram's superb account of the acreage development phenomenon testifies to how the public's passion to own a piece of the mountains threatens traditional land uses and landscape preservation. The firsthand account of the confrontation over Shell's assault on Prairie Bluff by Sid Marty, the award winning writer, is one of the best, most evocative explorations of the energy/landscape conflict in Alberta I have ever read. His critique of unbridled development is powerful. Less conspicuous, but equally important, is the essay's commentary on the public's failure to mobilize to try to save magnificent landscapes like Corner Mountain.

I.U.

THE BATTLE FOR THE FOOTHILLS[3]

- by Joshua Avram

Heading west out of Calgary, the cheek-by-jowl housing of the sprawling suburbs slowly gives way to the bald hills of the Municipal District of Rockyview and an amorphous community called Springbank. The roadsides are festooned with signs trumpeting the impending arrival of pre-planned, gated communities and acreage subdivisions. But one of the billboards stands out from the rest. It looks like a baseball scoreboard, green with white letters, and it reads: "Springbank 6, Nature 1." The scorekeeper is a curmudgeonly dairy farmer who has watched the land around him transform from open range into one of the toniest neighbourhoods in the Calgary region. "If I go outside my door, I've got subdivision all around me," gripes the farmer, who requests anonymity because he has already got enough trouble with his new neighbours. "I listen to bulldozers backing up every daylight hour. Beep, beep, beep. You hear it from miles away, and it will continue for three years."

Acreage development in the southern Alberta foothills is hardly new, but the pace and scale of construction going on today is unprecedented. Calgary is awash in new money, and there is a powerful urge among many young, urban professionals to get their piece of the foothills. It is partly baby boomer nostalgia for a "back to the land" lifestyle, partly an escape from the stresses of city life, partly a hunger for the ethos of the frontier, and partly an opportunistic grab for some of the hottest real estate in the country.

Inevitably, the foothills land rush of the late 1990s is creating economic and social conflict between old-timers and newcomers. "Go back five years," says the dairy farmer. "The adjoining piece here was a nice alfalfa field that belonged to a developer. I would crop that land and pay him. Then it all gets dug up and turned into homes. Two years later, you find this same developer in the council chamber talking about preserving agricultural land. It is a little odd." Odd perhaps, but then again, every newcomer to the foothills – from Sundre in the north to Crowsnest in the south – wants development to end with the installation of his own west-facing bay window.

"Everybody wants to get out of the city," says Shirley Gee, a security scheduling officer who fled Calgary, first for Bragg Creek and later for Turner Valley. She also has a five-acre place up in Bearspaw, just south of Cochrane. "It's closing in, especially at night," she relates. "You used to be able to drive out and it was totally black. Now you can look across and there's lights, lights, lights. It's almost like a commune, only spread apart." The urbanization of the foothills troubles Ms. Gee. "Like what they're doing to Bragg Creek. It's horrible. It's so commercialized, all the older people that used to live there are now selling."

The area west of Calgary is "almost like an acreage ghetto," quips Penny Dowswell, a media manager with the Calgary Board of Education. Ms. Dowswell owned four acres in Dewington before moving north to a quarter section in the Cremona area. "You had skiddoos roaring around in the winter (in Dewington) and I didn't move to the country for that."

Ms. Dowswell has a utopian view of rural life that does not involve fast, noisy machines carving trails on a snow-covered pasture. "There's something about the country," she says. "You can just walk out the door and see the mountains and the trees and the birds." The desire to commune with nature between commutes to the city is a growing phenomenon, according to Ms. Dowswell. "I think it's a shift in cultural beliefs," she says. "I can remember my mom talking about 4-H. My grandmother was a rural doctor and my grandfather was a minister. They told me all of these stories of the pioneer times. I thought they were wonderful.

"Today, we are going to work and trying to rack up all this money. I'm wanting more out of it than this," she says. "People are saying, 'Hold on here, I'm not so sure about this.'" But finding sanctuary is proving more and more difficult. "The more they subdivide, the further out I'll move," she says.

For ranchers like Roy Copithorne, such attitudes, however sincere, are proving to be a problem. "Almost everybody that moves out here comes with an attitude of 'Let me in and close the door, no more subdivision,'" says Mr. Copithorne, who works 10,000 acres 30 miles west of Calgary. "They don't feel that their moving out has any impact on the area, because 'It's just me and my family... We can't be causing any problem by being here. That's not our intention.'"

Mr. Copithorne thinks so-called "rurbanites" are nothing more than a generation three times removed from the farm seeking a shallow reconnection with their roots. "They don't want to return to the lifestyle," he insists. "They want to return to the memory. They want to be on an acreage out of the city, but they don't want to work it. They just want to be there and play on it."

Although such desires sound harmless, they are often accompanied by ill-informed environmental beliefs. "They question everything you do agriculturally," Mr. Copithorne says. "My cousin is taking a stand of poplar trees 35 to 40 years old and turning a bulldozer loose in it. They're really upset about him destroying the so-called 'forest.' They like looking out their window at a sea of trees. But because they're new to the area, they don't understand that historically, it was rangeland."

A blight of rurbanites has already forced Mr. Copithorne to move further away from Calgary. "We had 320 acres in Springbank we left strictly because of the dog problem," he explains. "You move 180 head of heifers down to your half section for pasture Tuesday morning at 5:30. Your neighbour phones to tell you your heifers just went past his place. You drive out and chase them back to the field and find seven panels of fence torn out because the acreage owners' dogs pushed the cows into that corner and the cows pushed through the fence to get away. Then what you have to do is go down and shoot the dog. We decided that rather than fight with the neighbours, it was easier to sell the property and move out." The half-section is now occupied by a subdivision.

The conflict between the ranching and acreage communities is as much economic as it is cultural, perhaps moreso. As the demand for residential real estate in the foothills grows, it is driving up the price of the land far beyond its normal agricultural value. Mr. Copithorne's neighbour, Art Froese, says escalating land values present a significant obstacle for ranchers looking to expand. "North of Bragg Creek, the land is very highly priced because there is very little for sale," he says. "Over the last 10 years, the price of land has gone from $150,000 a quarter, without streams, to $650,000. That's the order of magnitude." Hiesem Amery, an agent with Century 21 Bamber in

Calgary, confirms that Bragg Creek listings are hot. "If two quarter sections in Bragg Creek went on sale for $250,000, people would be killing each other for them. You don't have to see it. You'd just buy it."

But the buying frenzy is not limited to Bragg Creek 30 minutes southwest of the city. Carol Anne Kirkham is with Chief Mountain Realty Ltd. in the Crowsnest Pass, where features like nearby Waterton Park are drawing Calgarians in droves. The buyers are looking for vacation or retirement properties, or they are telecommuters who work from a home office or speculators anticipating a population, demand and price boom throughout the region. "Two-and three-acre parcels are well over $75,000 next to the park," she says. "It's nothing to get $80,000 for an unserviced five-acre parcel. And that was last year."

Even locales as far north as Sundre are feeling the foothills' rush. "Anything that's got trees, a creek, or a mountain view is gone," says Shirley McDonald, an agent with Re-Max Foothills Realty in Sundre. "We've just got a few of the poorer ones left, and they're selling for anywhere from $45,000 to $55,000 for five-acre parcels with no services." Sundre-area resident Bill Allen says the change has been dramatic. "Ten years ago you couldn't give a quarter away. Now they're lapping them up and it's driven the price sky high," he says. "I know there's places closer to the city, like Water Valley, where five, six acres are going for $65,000 to $75,000."

Surprisingly, Ms. McDonald says, many potential buyers are unaware of how hot the market is. "People still phone us from the city and they want a quarter section by the river with a mountain and trees. I usually finish it for them and say 'For $50,000.' They say, 'Oh, you've got one.' People have no idea." Bill Field, an agent with Realty World in Crowsnest Pass, hears the same requests. "People call me up and say, 'Bill, find me a quarter section in the mountains.' Well good luck! Three years ago you could have got one for $28,000. You can't get one for $50,000 today."

Mr. Copithorne says escalating land prices will drive cows off a pasture just as effectively as a developer's bulldozer. "In 1971, we bought a quarter section for $200 an acre. Today, if I wanted to sell it fast, I'd list is at $4,500 an acre," he says. "At $200 an acre, it was marginally feasible, because my cows were worth 55 cents a pound. The calves I sell this fall are worth a dollar. How do I justify staying agricultural in this area when the land value's gone up 20 times and all the input costs have doubled or quadrupled, and the cows have only doubled?"

Not all acreage owners are unsympathetic to the plight of ranchers. "We're taking prime land for agricultural use and putting houses on it," says an acreage purchaser from Calgary who recently settled on two acres in Springbank. "All these farmers are getting to the end of their careers wanting to give the stuff to the kids. But the kids don't want it and so they get rich by selling it. It's a conflict between old and new ways of life. If I was a younger guy who was a rancher I'd be quite upset."

Some rural residents are. "These guys around me, they all love me," scoffs the Springbank dairy farmer. "Why? Because I'm preserving agricultural land? No, because they like to have open country space around them. If I ever did what they do they would hate me. I would be a dirt bag, and then they would go to council and start squawking about preserving agricultural land."

The social changes brought on by subdivision are also awkward, he says. "Basically you've got no friends left. At a school function you'll dress differently. They say, 'I guess you get used to the smell of manure, hey.' Or your kid goes to play at Martha's house," he says. "It's a palace, you think Lougheed lives here or something. You don't even want to go up to the door because maybe you're wearing rubber boots. It sounds stupid, but people don't want to live where they feel they don't fit in at all."

The effect on local politics is similarly dramatic. "If it becomes me and 400 acreage people, obviously my vote becomes meaningless," says the farmer. "Not only that, but our political ambitions are different. I'm happy with a gravel road; they hate a gravel road. And how many schools do the 400 acreage owners need? Virtually every single issue the urban and rural people see it differently."

Further north, while driving into Water Valley, Hiesem Amery and fellow agent Chloe Cartwright of Century 21 Bamber in Calgary are discussing how ignorant many newcomers are to the basic necessities of life in the country. "People buying isolated quarter sections better know their stuff," says Ms. Cartwright. "If it doesn't have power, gas, or a well, those are substantial costs." The realtor used to conduct an acreage buyer's seminar every year. "One of the first questions I'd ask is, 'Did you grow up on a farm or acreage?'" If they are coming from the city, she says, they often do not know about providing for water, septic tanks, school buses, road clearance, and other such staples of rural life. Mr. Amery interrupts and points to a house trailer. "One or two acres with a house, $125,000," he says. "There's some stuff in there that's cheaper, but it doesn't have running water."

He pulls up to the Water Valley Saloon, stops, and chuckles at the hitching post out front. On Saturday nights, it is not unusual to see a couple of horses tied there awaiting owners who want to avoid the risk of a drunk driving charge. Walking into the saloon, he encounters a much less amusing sight. A bulletin board on the back wall of the lobby, covered with advertisements for local events and used cars or machinery, also features a couple of local realtors' business cards: both have several tacks jammed into their pictures.

"They're pretty cutthroat out here," reports Bill Allen, who is selling his acreage privately. "A lot of manipulation was done by certain realtors to make people pay more for land," agrees Anne Eskesen, who along with her husband Lyle runs Sundre Feed, Farm & Pet Supply Ltd. But local landowners were not unhappy when they found that "by selling five acres they could make a nice little tidy sum off it. This trend started to happen when it became the thing to find a corner acreage."

For long-time landowners, "many of whom are reaching their 70s and don't want to farm any more," she says, "it was a good time to sell." For others, "it was the source of bringing in additional revenue to renovate the house." Mr. Eskesen says the ensuing buying frenzy has driven farmers further north where land is cheaper. "All the young farmers can't make it," he says.

Realtor Kirkham says the same phenomenon is occurring in the southern foothills. "A lot of the people that have been selling are the older folks, or their children thinking they can get more money by selling," she explains. "For somebody coming in for a cow-calf operation or ranching, at these prices they couldn't make a living."

It is a situation that has received insufficient attention from agricultural operators, says Harvey Buckley, a rancher southwest of Cochrane and a member of Action for Agriculture, an organization dedicated to protection of the industry. "We are losing over three million acres (of agricultural land) every decade in western Canada to other uses. At the rate we're going, North America will be a net importer of food in 50 to 60 years," he says. "Unfortunately, farmers are very complacent for the most part and have not gotten the story out there."

Such complacency is making the job of preservation even more difficult. "It now appears in the next 15 years almost all deeded land within a three-hour drive of any urban major centre in Canada will either be compromised by urban or recreational development or will be under threat of same," says Larry Simpson, director of projects for the western and northern region of the Nature Conservancy in Calgary, a non-profit organization dedicated to private conservation in cooperation with landowners.

The difficulty, however, is that there are few people to complain. "For landowners, it's like winning a lottery," says Mr. Copithorne. "It's only the viable operations that are opposed to subdivision." Mr. Simpson agrees. "As the boomer population looks outward from their urban centres to buy land in traditional agricultural landscapes, they have the resources to pay three to five times its agricultural value. For landowners, it's a pretty powerful temptation."

The increasing number of such sales is leading to what Mr. Simpson calls "the Europeanization of North America. It's not just in Alberta and it's not just outside Calgary," he says. "It's estimated by the Utah Open Lands organization that some 60% of the ranches in the U.S. will change hands in the next 20 years. In Colorado alone, development now claims an acre of agricultural land every four minutes and 30,000 acres of wildlife habitat each year."

The glut of subdivisions in Colorado is most evident west of Denver, says Mr. Copithorne. "The acreages west of Denver extend for 100 miles," he says. "The sites with scenic rock ridges and big trees and water are the ones developed. The guy across the valley that almost had a good site has it subdivided for houses but nobody wants to live there. They've got an oversupply, but they're still asking too much money for them or they'd sell."

High prices have not slowed land sales in Montana according to Jim Peterson, executive vice-president of the Montana Stock Growers Association. "We are seeing some very large ranches that have been in families for many, many years put up for sale, and there's a lot of money chasing after them." He says one such ranch was recently sold to a California buyer for US$20 million. Atlanta media magnate and neo-environmentalist Ted Turner has accumulated hundreds of thousands of acres in Montana and 1.1 million acres in New Mexico, where he is the largest private land owner in the state.

Mr. Peterson says Montana ranchers face the same problems as their Alberta counterparts. "Traditional ranching and agriculture are in many cases not making enough money to be a viable economic enterprise. With land prices escalating, ranchers are looking at alternative sources of income, which may include subdivision and sale."

As in Alberta, the immigrants to rural Montana from the metropolises of the east and west coasts run the risk of undermining the rural life they are seeking. "One of the reasons states like Montana have enjoyed so much open space and wildlife has been because of the traditional ranch and agricultural values and lifestyle," says Mr. Peterson. "The things these people are looking for when they come to this state may not be here long term because of the change they bring with them."

Some observers think the complaints from farmers and ranchers are parochial and self-serving. Gordon Johnson has been a realtor in Pincher Creek since 1967, right in the heart of "a prime piece of the world. Those mountains are so close you can almost touch them," he says. "You could do almost anything here you could do at Banff except be a snob." Mr. Johnson accepts that agricultural land values are rising, but argues that the threat to the industry is overblown. "I don't think there's a shortage of land," he says. "We're going to have ranch lands here for a very long time."

Some acreage owners contend ranchers have no one but themselves to blame for the changes that are occurring in the rural landscape. "The people critical of acreage owners are the very same people who sell to acreage owners," says Ed Ryan, who owns eight acres five miles southeast of Lethbridge. "Then they turn around and make you feel like a pariah for moving into the country. It's so damn hypocritical. If you sell me a car, don't criticize me for buying the thing."

Mr. Ryan contends that in his area, acreage owners "have added much to the beauty of the countryside. We've got lots of nice trees around here. It's beautifully landscaped. I like to think that if this property had been left abandoned it would still be filled with weeds." Much of the subdivided land was of poor agricultural quality anyway, he adds. "Our land here was declared unfarmable. The guy who owned it said, 'Well I can put it up as an acreage and still make a handsome profit on it.'" If disappearing ranchland is a problem in other areas, says Mr. Ryan, "it's a problem of their own making. It is not one imposed on them by acreage owners. If they feel so adamant that the land should remain with another farmer, they should sell it to him at a lesser price."

While the growing rural-urban conflict has been noted by the provincial government, "we haven't thought we've reached a point where as the agricultural department we should propose that (subdivision) stop," says Mike Pearson, a policy analyst with Alberta Agriculture. Changes in the Municipal Government Act have made such interference even more unlikely. Prior to 1995, subdivision was dealt with by 10 regional planning commissions, part of whose mandate was to preserve agricultural land in all areas of the province. But as municipalities demanded increased control over planning, the government abandoned the commissions in 1994 in favour of general land use policies and voluntary organizations.

In effect, says Mr. Pearson, planning is now a municipal responsibility. "There are some general provincial directions. Provincial land use policies encourage municipalities to keep land uses compatible with one another and to try to identify areas where agriculture should be the primary land use. But it's pretty well up to the municipalities."

Municipalities have until September 1, 1998, to incorporate the province's policies into their own land-use bylaws and municipal development plans. Judy Mackenzie, a councillor for the MD of Rockyview, says public consultation has already shown that preservation of agricultural land is a priority. "We did a lot of open houses, and most people seemed to think we need to preserve agricultural land," she says. "You can't stop development, but you should look at where it happens."

Not all agricultural landowners agree that development is inevitable. "When I start talking about (slowing down development) they think I'm going to be strangling mothers that have more than two children," says the disgruntled Springbank dairy farmer. "People say, 'Well, you can't stop progress,' but the fact is you are going to stop growth at some point. I can't have an unlimited number of cows on my farm. I have to work within the economic and biological parameters. Do we want to wait until Calgary is Mexico City?"

However, the farmer expects his pleas to fall on deaf ears. He points to a recent article in the *Rockyview Times* where Langdon Councillor Jean Isley expressed her disdain for opponents of subdivision: "I'm fed up with people complaining about development," she said. "If you want to be a farmer, move."

If such attitudes prevail, it will likely mean that conservation solutions will have to come from the private sector. Mr. Simpson says one option for landowners is to enter into a conservation agreement with groups like the Nature Conservancy. "It's a voluntary agreement the landowner enters into to limit uses of the property which might harm the range lands or natural values." Restrictions can be placed on activities such as subdivision, drainage of wetlands or surface mining. The agreement is legally binding, and once registered, it runs with the title and binds all future owners of the land.

In other jurisdictions, governments have intervened directly to protect agricultural land. In Colorado, for instance, the state adopted a 1.5% land transfer tax with proceeds going to investment in agricultural conservation agreements. In British Columbia, the government created an agricultural land reserve by designating 5% of the province's land mass as exclusively agricultural. However, Mr. Simpson suspects such statist interventions would be rebuffed in Alberta. "It's not an option from what I can see, yet something needs to happen pretty quickly."

In many ways, says Mr. Copithorne, it is a dilemma that ranchers and farmers have inadvertently brought on themselves. "Because we have been as efficient as we are, we as North American agriculturalists are our own worst enemy," he says. "North America has always enjoyed a cheap food policy, and the product that we produce is only cheap in North America. People will not understand that agricultural land is important to protect until they go to Safeway and that product's not there. Because it doesn't become valuable until there's a shortage."

Editor's postscript: At least one municipal district in Alberta, the Municipal District of Ranchland No. 66, in the foothills west of Nanton, has tried to inoculate itself against country residential subdivisions. In the interest of preventing the fragmentation of agricultural and grazing lands in the Municipal District, the Land Use Bylaw lists "grouped country residential uses" as a prohibited use.

HEADLIGHTS AT THE GRIZZLY'S DEN[4]

- by Sid Marty

Mike Judd never flat-out asked me to go up on Corner Mountain to help stop the lions of progress from clawing a road up Prairie Bluff, a 7,000-foot-high plateau southwest of Pincher Creek. As a matter of fact, it was his former wife, Wendy, who had phoned me one evening in early November. I could hear kids yelling in the background as Wendy explained how Mike (a guide and outfitter at Beaver Mines), along with James Tweedy (sic) (a local cabinetmaker), had discovered Shell Canada bulldozers building a road up the bluff, which is Crown land. They were not surprised by the discovery. Mike and James are long-time members of the Alberta Wilderness Association, and the AWA had appealed to the Energy Resources Conservation Board (ERCB) in September to deny Shell a gas well drilling permit on the bluff.[5] Their request was denied, as usual.

Mike and James had done something that marked a first for the AWA: they went and asked the work crews to stop the destruction. The men shut down their machines, and Shell called the police. Mike was warned that he could be charged with "public mischief" if he got in Shell's way again. That's what they call civil disobedience in the province where I was raised.

Prairie Bluff is Crown land located in the Alberta Forest Reserve. Corner Mountain (7,396 feet) rises from the north end of the bluff, and Victoria Peak (8,400 feet) rises from the south end. These are part of the chain of front range mountains that form the magnificent skyline between Pincher Creek and Waterton Lakes. The range is interrupted by steep-sided canyons, the headwaters of trout streams, canyons that funnel the chinook wind out over the adjacent prairie. Every one of these valleys contains roads, pipelines and gas wells. These mountains were once part of Waterton Lakes National Park, and I wish to hell they still were, because then somebody would protect them. This is valuable winter range, always in short supply, for a herd of bighorn sheep. It is part of what biologists refer to as the Crown of the Continent Ecosystem, which takes in mountainous country from the Bob Marshall Wilderness of Montana north to the Crowsnest Pass. This expanse of land is vital for the survival of wolves, grizzly bears and other predators that move back and forth across both international and provincial boundaries. Since 1950, 290 miles of roads have been built in the surrounding mountain area, resulting in a 45 percent loss of elk habitat and a decline in elk and grizzly bear populations, according to the Castle-Crown Wilderness Coalition.

The summit of Prairie Bluff offers a panorama east over the prairies and south to Chief Mountain that is stunning to behold. The bluff is windswept, the plant cover is scarce and slow growing and the alpine habitat it encompasses is not only fragile but is itself a very small portion of the Alberta land mass. These surrounds of mountains and river valleys offer 120 rare species of vascular plants, three times more than are found in Banff National Park.

The government had approved Shell's application to bulldoze roads and drilling pads on the bluff, build a powerline and drill for gas even though the sites were in a high-altitude "Prime Protection Zone" described by the Department of Energy and Natural Resources as a zone to "preserve environmentally sensitive terrain and valuable aesthetic resources." It was an unjust decision that would finally provoke normally restrained and cautious conservationists into direct action against the developmentalists.

If you live on the east slopes of the Rockies, and you love the mountains, you are just bound to wind up wrestling with oil company behemoths, and these companies, always on a tag team with the provincial government, seldom lose the match. This perception is shared by a lot of people who care about publicly owned land. Mike Judd summed it up this way: "The government and the oil companies are the same thing." You might say the government is a company's major shareholder, since it stands to scoop up millions of dollars in royalties from natural gas sales. (How many millions I can't say. Alberta Energy wouldn't tell me what our government received from Shell. The information was confidential.)

I wasn't sure if I wanted to take another psychic mauling from these two giants just then. I told Wendy I would think it over. I went to bed and lay awake for a long time. I told myself it would be crazy to go and stand in front of a bulldozer driven by a perfect stranger. It's not a good way to get acquainted. But then I started thinking about other things.

I thought about dirt bikes, skiddoos and 4 x 4s running amok, about hunters chasing elk on opening day in the Porcupines that year in a 4 x 4 truck, shooting rifles out of the window. I thought about all those oil company roads, out there in the darkness, groping around in nearly every valley along the east slopes. Headlights at the grizzly's den.

"I wonder if Mike's asleep yet," I said out loud.

"Of course he's asleep," said Myrna. "Just like I was until you woke me up."

I phoned Mike and woke *him* up. It seemed nobody else was available at dawn to go up on the mountain. Mike was worried about going back there by himself. "I wouldn't ask anybody to go along on a thing like this," he said quietly.

"Well, I'll go with you. I'd like to try and write something about it for the funny papers, to at least let people know what's going on."

That's how Mike and I wound up at Shell's upper well site on Prairie Bluff in the chilly hours before dawn one November day. A narrow valley here leads to the summit of the bluff. I pulled my old Dodge Coronet off to one side. The venerable landship let out a long gurgling sigh of relief, and assumed a position of neglect.

"You don't think they'll sort of 'doze the landship off the mountain, do you?" I asked Mike as we pulled on our packs.

"They'd better not," he replied encouragingly.

The well site, flooded with artificial light, was full of men and equipment. There were several large bulldozers, and a number of Rota-screw rock-drilling machines ready to screw the earth. Men were scampering around their rumbling charges looking efficient and keen to get on with the legalized assault on Prairie Bluff. Down below us, near the company field office, we could hear the whine of helicopter engines warming up.

The first light of dawn crept arthritically across the vast prairie below the mountain foot. It gleamed on the towering smokestack of a gas plant a few miles to the east. Fireflies of light streamed down the roads that radiated out from the plant like the glimmering tentacles of an octopus. A company supervisor came roaring up the road in an expensive-looking car. He eyed us suspiciously, slapped leather and drew his cellular phone or whatever. But he didn't try to stop us as we started walking up the mountainside.

"He'll be calling the Mounties, I imagine," said Mike.

We walked in silence for a while. I was feeling queasy about taking on the law. As a former park warden, I had been a bona fide peace officer, and I'd stood back to back with the Mounties on occasion to fight off platoons of rioting drunks in the park campgrounds. That was many dead Tories ago, but I still felt strange, as if walking up the trail into willful exile. Both of us live on a shoestring budget. As a guide and outfitter, Mike cannot afford to place himself in a confrontation with government agencies that regulate where and how he does business. But those same agencies had presided over the destruction of far too much of the wild terrain he needs to stay in business. Tourists are funny. They don't enjoy trail riding through clear cuts, seismic lines, powerlines and gas well sites. Underneath Mike's calm exterior, he was starting to do a slow burn that would lead to an explosion.

I glanced at him. "How many people will be coming out to protest this thing?"

Mike grinned. "There'll be at least four here from the AWA by 10:30."

Four!

Surveys have shown that 80 percent of Albertans think our wildlife should be conserved. Wildlife needs wildlands. There are thousands of people in this province whose incomes depend on the existence of wildlands: photographers, mountain guides, biologists, outdoors writers – the entire tourism industry. What about all the storekeepers who sell cross-country skis, hunting and fishing equipment? What about the thousands of backpackers, hunters, skiers, the rangers, wardens, professors of environmental science – an endless list of potential defenders, and we had five protesters and one writer. No wonder we're being eaten alive by "progress."

It was 7:30. Mike was going to try to hold off the lions for three hours. He had only one card to play.

A cold wind, a hard wind that whipped particles of shale against our faces like buckshot, was coming down the little valley. I could see a narrow white scar leading up the mountainside, an old trail partially drifted in with shale. It was built twenty years ago to install a radio tower on Corner Mountain. The summit of the peak, at 7,400 feet, was up there beyond the red rock bands of Prairie Bluff, which forms a high plateau at its base. Up above us on the summit of the bluff were the two drill sites. The bulldozers would widen this road to get to the sites, towing the rock drills behind them. This old trail could not be made safe for permanent use, according to Shell Canada, but opening it temporarily meant they could start work on their drill sites right away. Once on top, the crews would begin cutting their way through the fragile meadow, going deep enough to grade out a level spot on the steep slope. They would have to push aside a thin soil cover in the Prime Protection Zone that had taken 10,000 years to accumulate. One operator would work his 'dozer down from the summit, following Bar Creek to the south, building a section of the new road, and eventually meet up with the one Mike and James had confronted on November 1. That 'dozer was now stopped by a rock spur that would have to be blasted out with high explosive.

When Shell were finished building on Prairie Bluff, the ram pastures there would be an industrialized landscape. There would be a new road up the tiny perfect valley of Bar Creek, blasted through rock spurs on forty-five degree slopes. Shell promised to install a gate on the new road to deter 4 x 4s.

Gas wells on the mountain tops! At least Shell does not tolerate any litter around their gas wells, and they keep their pipe mazes well painted. Royal purple was one of their favourite colours at that time.

The wind that first morning was gusting hard enough that we had to brace both feet to stay upright. We got down behind a stand of dwarflike bullpine, and instantly found ourselves in a little pocket of calm in which to wait for the bulldozers. While the sun's first rays arched over the prairie, we sat mulling over the events that had brought the AWA to this confrontation with a multinational corporation. Of the 4,500 drilling permits requested by the industry so far that year, these two permits were the only ones that conservationists had contested, because they were in the Prime Protection Zone.

Now one of the things that deeply troubles some Albertans is that permits like Shell's can be issued without a full environmental impact assessment (EIA) being undertaken. That is a survey of effects not only on recreation but on many other features, including wildlife, soils and plants as well as the social and economic effects of development. The AWA had been attempting for over a year, without success, to get the Department of the Environment to do an EIA of Prairie Bluff. Such an assessment is admittedly expensive, but the government does have employees and resources for doing such work, and the oil companies should help to foot the bill for it, in my view. After all, the gas that Shell wanted to produce on this publicly owned mountain was destined for export. The main beneficiary, after royalties are paid, was a foreign-owned corporation, Shell Canada, that generated $5.3 billion in sales and operating revenue in 1985. Shell hoped to recover 20 billion cubic feet of gas, worth about $300 million, from these new wells. Even a million dollars for an EIA would not be out of line. To Shell's credit, they employed a biologist to do an inventory on the bighorn sheep population on the bluff (at a cost of $75,000). Still, in the United States, an EIA would be mandatory in a case like this. Our government has no policy requirement that EIAs be done for any project, even when they take place at the source of our drinking water supplies.

The bluff provided vital winter range for bighorn sheep habitat in an area often cited as the best bighorn sheep habitat in the world. At a September hearing, Shell's biologist testified that the development shouldn't interfere unduly with bighorn sheep use of the bluff for winter range and mating activity. Shell would restrict its activities to "windows" of use. They had until November 30 to complete their construction. Then the bighorns could return until drilling activity began for the May 1 - November 30 window. Shell was trying to show good will to the bighorns. But wait a moment; weren't there other mammals living on the bluff – deer, bears, wolverines etc.? What about them? And what about the rights of Alberta citizens who don't enjoy hiking, riding and experiencing wildlife right next to an industrial development, on publicly owned land? Such concerns went unheeded.

As we sat sipping our coffee that morning, we could see a huddle of men down below us obviously confused over what to do about the two "crazies" up on the mountain trail. It was then I spotted the police car wending its way up the switchbacks down below. Sure enough, one of the bulldozers idling at the foot of the trail finally came to life with a metallic scream and started grinding its way up the steep slope toward us. The cruiser pulled into the parking lot below. We stepped out on the trail in the dim morning light. The huge orange machine crawled closer, closer. I could see the operator huddled over his controls as the shiny blade lifted higher, the cat tilted up, the blade hung suspended in mid-air.

"Just stand your ground," ordered Mike.

"Uh, Mike. Remember I'm here as a journalist."

Mike laughed unpleasantly. He glared down at the well site below us, and muttered an outdoorsy saying or two. The blade slammed down into the gravelly trail. There was a hummock of gravel and a shallow ditch that was supposed to deter dirt bikes from roaring up the mountainside: it never has. The big blade bit into it, and shoved the pile toward us, flattening it out. A few stones rolled to a stop in front of us. I made like a reporter and took some pictures. The operator stopped the machine and stared at us. I thought, here we go. Now he's going to blade us right off the mountain.

Instead, he just shrugged his shoulders and grinned as if to say, "I just work here," and sat back in his seat to wait for instructions from his foreman.

"Keep him here," said Mike suddenly. "I'm going down there to talk to them."

I swallowed nervously. "Why, what's up?"

He hurried over to talk to the operator. He was going to play his one card.

"Have you got a permit to go up here?" he shouted over the engine noise.

The operator shrugged again. He had his orders.

Mike turned to me. "They need a letter signed by Art Evans to go up this trail."

I stared at him, catching his drift. Evans was the Chief Forest Ranger for the area, based in Blairmore, Alberta, some miles to the north. Without a letter from him Shell had no legal right to begin work here, and any citizen certainly had a right to ask them to stop.

Mike went tearing down the hill to talk to Shell's foreman. I went over to talk to the operator, get his point of view. He said it's too bad that we have to do this, but I just work here, or words to that effect. The policeman was standing by his cruiser just then, watching events unfold. Everything was too well orchestrated, like a badly scripted play. In a few minutes Mike came panting back up the trail.

"They don't have it!" he cried, elated. "Can you believe it! They never even bothered to get that letter. We've got 'em."

The victory was short-lived. A forestry truck soon appeared on the scene. We watched as the Mountie went to confer with the new arrivals, a forest land use officer from Calgary and a forestry officer from the Blairmore office.

"They're cooking up that letter right now," I warned Mike.

"They can't do that!"

I didn't say anything. I had some experience in these matters. The forest officers could act for Evans if he gave the word by radio, and I didn't see how anyone could do much about it.

We stood in front of the 'dozer, the hard wind blowing at our backs, getting down the back of our necks, getting closer to the bone. Pretty soon a small knot of men started climbing up toward us. One of them clutched a scrap of paper in his hand. It's always amazed me how much damage can be done to entire mountain ranges just with a few words scribbled by a ballpoint pen.

I looked at Mike, a tough, medium-sized man in a black wool cap, a pack on his back, worn climbing boots on his feet. He is a horseman, and has the right build for the trade. Behind him the mountain loomed, a threatened giant, an elephant guarded by a pair of ants. By 9 a.m. that morning, it would be all over but the shouting. We would have to stand back and watch as the 'dozers bit deep into the side of Prairie Bluff and the falling rocks went rattling down the shale slope.

We left the mountain shortly afterward. By the time we returned later that morning with Vivian Pharis, president of the AWA, executive director Diane Pachal and two other members, the old trail had become a road once more. The 'dozers were already starting work on the drilling pads on the summit of the bluff.

That was the beginning of a week-long confrontation between the AWA and Shell that only subsided when Shell got an injunction to keep the AWA and everybody else out of the working area. The AWA managed to slow down some of the work before the injunction came into effect. Members would walk in front of the 'dozers carrying picket signs. News media from Calgary and Lethbridge filmed some of the action. So did Shell.

To some of the operators, the pickets must have seemed like members of a union, manning a line that they, as fellow workers, should not cross. As one man said to Mike Judd, "That's okay, you have a job to do I guess." He was nonplussed when Mike explained, "Hey, I don't get paid to do this." It was puzzling for the workers. I think they started resenting the protesters after a while for expecting them somehow to be better than they were able to be, to suddenly develop the elegant principles of conservationists when they were working men with families to feed and not predisposed by habit or inclination to spend time informing themselves on complex environmental issues. They had to make a living. Here you are, working away, proud of your skills at cat skinning or blasting, and those bozos keep waving their signs at you – "Save the Bluff" – as if what you are doing is somehow evil. This makes you irritable, and on two occasions the protesters felt the earth move before an overzealous operator backed off. But it was obvious that Shell was determined not to make martyrs of the opposition.

There were some bizarre moments that week. The sight of news reporters climbing up a mountainside in shiny patent-leather shoes and trench coats was good for comic relief. Mike Nikotuk from the Canadian Broadcasting Corporation conducted one of the strangest interviews I've seen in a while. By that time there were five protesters on the bluff, and more were on the way. Three bulldozers were blading up the heather at one site. Three hundred yards to the north, crews had already stripped the topsoil for another pad just below the summit of Corner Mountain after pushing a trail through the forest to get there. I'm not sure that Nikotuk grasped the terrible damage that had been done to the fragile soils of the mountain. He gathered the little knot of protesters together and began firing questions at them. Nikotuk, recently arrived from the west coast, wanted to do a new angle on the AWA. His view seemed to be: If this were Vancouver, there'd be thousands of demonstrators here. Obviously the people of this province don't agree with your position, so why do you bother protesting?

Nikotuk hit a sore point. Diane Pachal pointed out that there are 2,500 members in the AWA alone. I couldn't help wondering myself where the hell they all were right then. They couldn't all be working, out of the country or having babies at that moment. Of course, this is the problem with leaving environmental police work to volunteer organizations. Somebody working for an oil company as an in-house "social anarchist" (i.e., Environmental Co-ordinator) was making forty to fifty thousand dollars a year from environmental issues. They need an aroused citizenry to keep them employed. It is the AWA and other organizations that arouse the citizenry who don't have time to track every environmental outrage going on in this province on any given day. Those who try to stop developments like this have to take time off work and pay their own expenses in order to do the work that well-paid civil servants and politicians are supposed to be doing for us. Eventually, this cripples you both financially and psychically. But Diane Pachal probably gave the best answer that day on the bluff. She swept her arm around to indicate the beautiful plateau, the prairies to the east, the towering wall of Victoria Peak behind us. "I believe there are thousands of people who support our stand here. But even if I was the only person in Alberta who cared about this beautiful landscape, I would still be here. These mountains demand it of us."

My thoughts went back to the first day of the picket, when Mike and I had been alone. We stood in the howling chinook wind that blew down from the summit of the bluff. Constable Foster of the RCMP told Mike he would have to stand aside and let the bulldozers come through. Backing him up were the two Forest Service officers, Shell's foreman and several other employees. I exchanged ironic looks with the forest rangers, who were known to me. No words were said, but I knew that, hard as it might be on Bertie the Beaver, he was going to have to struggle along without yours truly.

"Where's the letter from the Chief Ranger?" demanded Mike.

One of the rangers produced a sheet of paper with a handwritten note, and handed it to Mike.

Mike glanced at it and handed it back. "That's not good enough. That's not signed by Art Evans."

Constable Foster, a neighbour of Mike's and a man he likes and respects, said quietly, "Mike, it is good enough. And I'm telling you, you'll have to accept it."

As they spoke, the wind howled in stronger protest. It struck Mike in the back and threw him forward, as if urging him into the fray. I caught him by one arm to stop him from flying into the opposition. We were all holding on to our hats, our opponents leaning forward and moving their legs as if on a treadmill to hold their ground, while Mike and I had to lean backwards and shuffle uphill. One of the men tried to explain how Shell was going to repair all the damage they would do on the mountain when the gas wells on the bluff went dry. I believe he was sincere, but the wind blew the words back down his throat.

That's when Mike finally let go some of the bitterness that had been building up in him over the last decade of exploitation in these mountains. He had been to so many east slope policy information sessions and hearings, had swallowed so many paragraphs of bafflegab promises and buzzwords from government flak-catchers and industry mouthpieces while the bulldozers and chainsaws did their deadly work, that he was full to the brim with it all. Now he spat it all back up, and cut to the heart of the issue.

"Look at that road behind us!" he yelled over the roar of the wind. "That was done twenty years ago and the scar will last hundreds of years. You'll be gone from here in twenty years, and this is what we'll be left with! This is the best bighorn habitat in the world. You'll never put it back, you'll never make Bar Creek like it was. You can't do it! You –"

Mike's voice was partially drowned out as the wind hit him again, throwing him forward. He was yelling some pornographic descriptions about his opponent's lineage. I caught him again. The wind pushed us all down the steep slope as if we were sailors plunging through a gale.

Constable Foster tried to comfort Mike. He put his arm around him, his other hand anchoring his hat. "I know how you feel –"

Mike shrugged off the gesture. The wind spun him round and nearly knocked him to his knees. He said what we all say in that situation. "No you don't! You don't know how I feel." Mike was the loneliest thing in the world at that moment, except for the mountain brooding over his shoulder like a beast at bay. "It's greed!" he yelled. "That's all it is. After all the bullshit, it just gets down to one thing; just greed!"

The other men looked away, embarrassed. Albertans hate making a scene while they are trying to eat. Whether Shell realized it or not, it was the brave notion of the Prime Protection Zone, it was the very idea that something should be allowed to exist that wasn't generating money – that idea was what it had come to eat that day.

The bulldozers were lining up below now. Tons of iron and steel, thousands of horsepower. The helicopter was lifting off its pad. One fierce-hearted man stood in front of them for a final moment, and Prairie Bluff, sending its winds in protest over the ram pastures and the coloured shale, over the weathered little fir trees and the tough wisps of flower stalks, awaited its fate.

Mike stepped aside and climbed up the mountainside. He sat down, fighting back bitter tears. In a few minutes he would apologize to those minions of the law and progress for his rough language. The people who should have been there to get the benefit of his rage were safely hidden behind steel towers in Calgary and Edmonton, Toronto and London, or in the helicopter that would soon be circling overhead, watching our every move.

High above Prairie Bluff, on a ridge of Victoria Peak, there is a dream bed: a circle of stones, a lichen-covered cairn. For 12,000 years, maybe longer, young warriors came to this place to gaze over the prairie and stare at two great spiritual landmarks: Chief Mountain to the south, Crowsnest Mountain to the north. They fasted and prayed for a spirit guardian to visit them in visions, to grant them the secret of its power. They prayed mainly for courage, so they could flourish in a dangerous world armed with stone knives and spears, long before the bow. Often the spirit guardian took the shape of a bird or animal: an eagle, an elk, a grizzly bear. They would be directed to obtain an eagle feather, an elk tooth, a bear claw – some such totem of the spirit's power to keep in their medicine bundle. This eight-foot circle of lichen-covered stones, sunk into the mountain earth, is the memorial of all those generations. That is the total impact of the ancients on this piece of mountain earth.

That day, a magical thing had happened below the dream bed. Huge roaring beasts the colour of paintbrush flowers in the summer were devouring the earth of Prairie Bluff, sending up clouds of dust and smoke. They had only been there for a few hours, but they had transformed the mountain in those few hours, forever.

Goodbye, Prairie Bluff.

CHAPTER 3

PETROLEUM BOOM, ENVIRONMENT BUST?

Chinook

In the lodges of my people
Time was heavy... Winter's fleeting
Was too slow. The hills, like steeples
Of the village church below
Lay inches deep with winter snow.

Restless are the spirits of my people,
Chained by North wind's icy grip,
Too chill to hold, yet not so bold
To venture in the path
And face the last of Winter's wrath.

Beyond the dawn, and over the hill
Came sweet Chinook, young child of Spring,
Her breath soft-scented of the sage
That grows so old and sere with age;
The gift of life eternal its heritage.

Out of the lodges came my people
To live, to breathe again,
And like the sage, once dull and grey
To grow with Spring into the day
When winter's cares will wear away.

- by Leonora Hayden-McDowell[1]

INTRODUCTION

"Happy days are here again."

- Greg Noval, President, Canadian 88 Energy Corporation[2]

Indeed, they are. The 1990s have been very lucrative for Alberta's oilpatch. Like the winds of southern Alberta, the signs of the boom cannot be missed. In 1997, a record number of wells was drilled in the province. Even more holes could have been poked into the earth. A shortage of drilling rigs forced some producers to postpone drilling programs and encouraged drilling contractors to expand the size of the drilling fleet. The sale of oil and gas drilling rights, a barometer of industry optimism, also set records. Land sales poured more than $1.15 billion into the provincial treasury in 1997. In 1996, nearly 69,000 workers earned their pay cheques in either the upstream production or the service segments of the oil and gas industry.

Despite a severe slump in oil prices from more than $20 (US) a barrel in 1997 to less than $13 (US) a barrel in the spring of 1998, exploration and development remains impressive. To be sure, fewer wells will be drilled in 1998 than in the record year of 1997 but the level of activity remains high. In light of the slump, the Petroleum Services Association revised its 1998 well completion forecast downward. Still, 1998 promises to be the second best year in the history of the industry. Nearly 10,000 wells will be drilled in the province – well short of the approximately 11,500 drilled in 1997 – but more than at any other time in Alberta.

The happiness this boom generates in Calgary's corporate towers is not to be found in the more modest digs of Alberta's environmental groups.

Worry is more likely to be worn on their members' faces – worry that government insensitivity and intransigence may have made it too late to spare any ecologically significant area of Alberta from the probes of oil and gas rigs. Although the clash between the energy boom and sustaining our environmental heritage is being played out all over Alberta, this section directs most of its attention to events in the southern limits of the Rockies and Foothills.

We start in the Whaleback – one of the largest undisturbed tracts of a Rocky Mountain montane landscape left in Alberta. For Amoco Canada, the Whaleback is prized for the possibility that billions of cubic metres of gas are trapped thousands of metres below the surface. For environmentalists, landscape preservation must trump the area's energy potential. The first Whaleback essay introduces us to the various arguments which the Energy Resource Conservation Board (now the Energy Utilities Board) considered when they looked at Amoco's application to drill an exploratory well. The decision to reject Amoco's application was a rare victory for environmentalists before the board. Subsequent material suggests that the 1994 victory may be ephemeral. Despite provincial studies acknowledging the Whaleback's ecological significance, it seems likely that the Minister of Environmental Protection will decide that the Whaleback should host a variety of industrial uses – including petroleum exploration.

From there we journey to the Quirk Creek oil and gas lease in Kananaskis country southwest of Calgary. There, the provincial government's insistence that, when it comes to oil and gas, "the only good lease is a drilled and developed lease" has frustrated the efforts of Husky Oil to protect a portion of a foothills watershed from development.

I.U.

—THE WHALEBACK—

"There is an acute shortage of intact montane areas in Alberta. There is no shortage of oil and gas wells."

- Harvey Locke, Canadian Parks and Wilderness Society[3]

My family and I discovered the Whaleback by accident two years after our move to Edmonton. When you live in Alberta's capital and parents, siblings, and cousins live in the West Kootenays of British Columbia you are always looking for shortcuts to shave some time off of the twelve hour drive home (especially when your passengers include a five year old, a two year old, and a yellow labrador pup!). When we were home for Christmas in 1988 Jimmy Lorman, the manager of Nelson's ski hill and someone who regularly made pilgrimages to Calgary to see family, told me to try Highway 22 north from Lundbreck. "The scenery's great, far better than anything on Highway 2," he said, "and you'll save yourself at least half an hour." Jimmy was right on both counts (although the times he claimed for the Nelson to Calgary run are still beyond my reach). But, having driven through the montane landscape which straddles Highway 22 between Chain Lakes and the Lundbreck turnoff the time saved really does not matter. We always take that route now because the scenery, as you'll sense from the beginning of Eric Bailey's essay, is stunning.

The conflict over the future of the Whaleback's relatively undisturbed montane landscape symbolizes well the crux of the argument between mainstream environmentalists like Harvey Locke and provincial cabinet ministers like Ty Lund. The Harvey Lockes of Alberta's environmental movement are not the arch-enemies of resource development. They simply don't believe that the oil and gas industry should shoot seismic lines and poke holes into every square inch of the province. For the Ty Lunds of the Alberta government, this view is extreme and radical – definitely unAlbertan.

The conflict between Amoco and a coalition of environmentalists and local residents, retold in Eric Bailey's essay, led Andrew Nikiforuk to label the Whaleback "the energy equivalent of forestry's Clayoquot Sound."[4] This epitaph was probably premature. While the decision Alberta energy regulators made in 1994 about Amoco's application to drill an exploratory well in the Whaleback affected the energy industry's behaviour, the decision was just the first act of this play. Whether the Whaleback really warrants the Clayoquot comparison hinges more on how the public reacts to the treatment this Rocky Mountain landscape receives from the provincial government's Special Places initiative. The section ends with the trials and tribulations of the Whaleback Local Committee and the Committee's thoughts on whether or not oil and gas activity is compatible with the preservation goals of Alberta's Special Places policy.

I.U.

SAVING THE WHALE:

THE ERCB SAYS "NO" TO AMOCO– FOR NOW[6]

- by Eric A. Bailey

"The bad news is that there is a very vocal group claiming to speak on behalf of Albertans, against reasonable development of Canada's resources. They represented us at Amoco's Whaleback hearing...

- Dave Newman, Chairman, Amoco Canada[5]

Whaleback Ridge, 140 kilometres south of Calgary, breaches the plains at the feet of the Canadian Rockies just north of the Oldman River, piling up against the mountains with other hills and ridges like waves suddenly frozen in their final rush for the shore.

There, you will find things of the mountains: Engelmann spruce and Douglas fir in cool stands of coniferous forest, elk, goldenmantled ground squirrel and the dipper. There, you will find things of the plains: aspen poplar stunted and trembling in the chinook winds that repeatedly rake the ridge, balsam poplar in low places, prairie grassland, Richardson's ground squirrel and Nuttall's cottontail. There too, you will find such rare and uncommon species as the golden eagle, prairie falcon and upland sandpiper.

In portions of the Whaleback, you will find things of man: tame grass, herds of range cattle, many kilometres of barbed-wire fences, ranch roads, cow paths, horse trails and the tracks of countless pickup trucks and all-terrain vehicles.

The fabled western sky vaults above it all and beneath it lies a geological anomaly that may contain a really big, very valuable deposit of natural gas. But what you won't find there is a single oil or gas well. Not now. Maybe not ever. This is the other anomaly at Whaleback Ridge: gas but no gas well.

There's Gold in Them Hills

Looking at old seismic exploration data and the results of newer work they conducted on two seismic lines, geologists at Amoco Canada Petroleum Company Ltd. have found a geologic structural anomaly lying beneath the Whaleback that they feel has enough potential as a natural gas prospect to warrant drilling a test well.

According to Vince Rodych, external affairs manager for Amoco in Calgary, the company paid the provincial government $1.6 million of leases on the area's mineral rights and quickly applied to the Energy Resources Conservation Board (ERCB) in 1993 for a licence to drill an exploratory well on its lease. Amoco stated that this was an exploratory well to confirm its interpretation of the seismic data, test for the presence of natural gas and determine how much gas might be there.

Amoco calculates that its chances of finding gas with a test well are something less than 40 per cent. Even if gas is there, it may not be there in a large enough quantity to be worth producing. Nevertheless, the gamble is worth it. Amoco says gas reserves under the Whaleback could amount to more than 42 billion cubic metres, or about two per cent of the known remaining gas reserves in the province.[7] If it found such quantities of gas, the company would develop another 20 wells in the area, the ERCB permitting. Amoco suggests that developing that gas would contribute $10.2 billion to Alberta's gross domestic product and generate 1,500 jobs each year for 20 years. The company figures government revenues from the project would top $2 billion including more than $900 million in royalties.

Amoco points out that Alberta's gas production base is declining at a time when domestic and export markets continue to grow. Gas exports to the United States have doubled in the last decade. The Alberta government now collects more royalties from gas than oil.

The ERCB, which conducted a 10-day public hearing in late May on Amoco's application, released its decision on September 8, 1994. In the face of all this rosy economic data, the ERCB politely said, "No."

Blame It on the Montane

The ERCB says in its decision that it agrees a test well would confirm the extent of the Whaleback's gas resources. It also agrees that natural gas exploration and development is important to Alberta's economy. In its judgement, however, Amoco's application failed to serve "the overall public interest," primarily because of the impact development could have on the "ecological, recreational and aesthetic value" of the Whaleback.

Local residents, environmental groups and others who opposed Amoco's application brought that value to the attention of the ERCB. The Hunter Creek Coalition spoke for many local residents, in particular for seven families who live around the Whaleback. The Whaleback Coalition, another key intervener formed specifically to fight Amoco's application, represented the Alberta Wilderness Association (AWA) and the Canadian Parks and Wilderness Association (sic) (CPAWS).

The ecological value of the Whaleback figured prominently in the arguments against Amoco's application at the ERCB hearings.

Many ecologists and naturalists agree that the Whaleback is the largest undisturbed parcel of the montane ecological zone left in Canada. The things that live in an ecological zone define it. The montane zone contains a mixture of grassland, woodland and wetland plants and animals, some typical of mountain zones like the alpine and subalpine, some typical of plains zones like the boreal forest, parkland and prairie. This unique mixture of plains and mountain types that forms the montane has established itself on ridges like the Whaleback in the eastern slopes.

Large areas of montane zone also stretch, like long fingers, into the Rockies along wide river valleys such as the Bow, North Saskatchewan and Athabasca. Banff National Park, the Kootenay Plains Ecological Reserve and Jasper National Park contain significant pieces of the montane, but little of it remains undisturbed. Highways, railways, hotels, resorts, campgrounds, picnic areas, gravel pits, power lines, pipelines, dams and reservoirs, industries, and so on, including the townsites of Banff, Jasper and Canmore, all lie in the montane zone.

Debate at the hearings centred on three issues: how much damage wells and roads would do, whether development had to respect critical wildlife areas already zoned under another planning mechanism, and what role the proposed Special Places 2000 policy should play in deciding the fate of the Whaleback.

A Well, More Wells and Some Roads

As the company's application to drill a test well would certainly lead to further development if the test results proved positive, the nature of future development became a focus of debate.

Amoco presented three scenarios for development of the full field. The first called for drilling 20 single wells on single pads scattered throughout the area. The second grouped two wells on each of only 10 pads. The third grouped 20 wells on five pads. Amoco advanced the third option, employing "extended reach" drilling technology, as its preferred option. Option three entailed the least disturbance to the environment from drilling and operating the wells, and from related developments such as roads and pipelines. The company also proposed to locate any processing facilities outside the Whaleback.

Opponents of the application said drilling the Whaleback would open up the area to more public access than it had ever seen. They argued that the Whaleback's montane ecosystem had survived over the years because no roads had reached the heart of the ridge. Wells and roads went together. Opponents feared the full development of 20 or more wells would lay the whole area open to uncontrolled access.

Hikers, horseback riders and off-road vehicle users have access to the Whaleback now, but the level of recreation remains low. Off-road vehicles already leave tell-tale scars on the landscape. The AWA would like to see off-road vehicle access curtailed. Local residents worry that more roads will lead to increased traffic and vandalism, dust, gates left open, cut fences, rustling and random or accidental shooting of cattle. The ERCB accepted these arguments, noting that a "sound access control plan, consistent with the Integrated Resource Plan for the area, should be developed before any disturbance is allowed to take place."

IRPs: Guidelines or Rules?

Amoco and its opponents argued about the importance of the Livingstone-Porcupine Integrated Resource Plan (IRP) that the provincial government developed with public consultation as part of the Eastern Slopes Policy. The IRP designates most of the Whaleback Ridge area as a critical wildlife zone. The area is essential wintering ground for about 1,000 elk and may be important to local populations of black and grizzly bear. Interveners opposed to the application felt the ERCB should be bound by the provisions of the IRP. Amoco, on the other hand, said that habitat loss would be minimal and could be mitigated, and argued that the IRP was only a guideline that did not bind the ERCB.

A Special Place – Maybe

Finally, the argument came down to the question of the value of the Whaleback as a complete ecosystem, not just as winter range for elk or spring habitat for bears. The AWA has lobbied the provincial government to formally protect the area as a "wildland recreation area" for over a decade. The government has set aside the Upper Bob Creek Ecological Reserve. But at 67 square kilometres, the reserve is only a small portion of the 236 square kilometres the AWA has in mind. The reserve does not cover enough of the ridge to sustain the ecosystem. The Whaleback Coalition favours withdrawing the entire Whaleback from resource development.

Opponents of the application cited the report of the Special Places 2000 advisory committee, which proposed a policy to the provincial government for protecting significant wild lands in the province. Although it does not mention the Whaleback by name, the policy document does mention the need to protect an example of montane ecosystems. The proposed policy went to the provincial cabinet in February 1994, but cabinet has not yet decided its fate.

Environmentalists, and planners with Alberta Environmental Protection, see Special Places 2000 as a means to "complete" Alberta's provincial park system by including representative samples of all the major ecosystems in the province. At the hearing, the Whaleback Coalition argued that the Whaleback would be a prime candidate for inclusion in Special Places 2000.

In denying Amoco's application, the ERCB wrote that approving the application "could significantly affect the area's surface values before the Special Places 2000 program has had an opportunity to evaluate the importance of the area in a provincial context."

The opposition had won.

More Policy, Please

"We were dumfounded," says Judy Nelson, a member of the Hunter Creek Coalition, who lives about seven kilometres southeast of the proposed well. Her husband's family has ranched in the Whaleback for three generations. She says the coalition talked to eight law firms before it found a lawyer. Even he told the coalition that its chances were small. "The other seven said we shouldn't even bother opposing the well," says Nelson.

The "no" ruling surprised members of the Whaleback Coalition as well. But some members criticize the provincial government's role in the affair, especially in failing to move Special Places 2000 forward.

Wendy Francis, a spokesperson for CPAWS says: "The Whaleback has served an important role in showing the provincial government there are land use conflicts that can't be resolved until a policy is in place. The ERCB decision has also shown the oil and gas industry why it's to their advantage to have a program like Special Places 2000 in place to establish some ground rules."

Vince Rodych admits Amoco was "both surprised and disappointed. We spent time and money entering into a forum we thought could deal with our application only to find out other policy issues need to be resolved first."

Rodych also expresses concern that the ERCB hearing process "opens up a forum for very vocal groups of small numbers of people who I don't think represent the true public interest. I think the public is in favour of a compromise in this situation and I think the public has been given the wrong impression of the concern for the environment in our industry."

If No Well, Why a Lease?

Speaking for the Whaleback Coalition, Francis says: "The whole system of mineral leasing is flawed from the beginning." According to Rodych, Amoco agrees that the system didn't work in this situation.

Francis contends that Alberta Environmental Protection could have refused to grant Amoco a lease of the mineral rights to the Whaleback. But, she says, "They have never said no. The history of the Conservative government is that they never say no. They leave it to the public interest groups to fight those battles and it becomes the responsibility of the ERCB and the Natural Resources Conservation Board (NRCB) to make these decisions instead."

The Crown Mineral Disposition Review Committee, an inter-departmental body of the provincial government, decides whether to lease mineral rights and under what conditions. Alberta Environmental Protection chairs that committee. Alberta Energy, which administers mineral rights, reports that no one on the committee suggested withholding the mineral rights to the Whaleback. No one mentioned the pending Special Places 2000 policy nor the priority of the Whaleback as a Special Places candidate. The committee discussed nothing more stringent than the integrated resource plan for the area.

Subsequently, Amoco leased those rights for $1.6 million.

The government had another chance to deny Amoco access to the Whaleback when it received the company's application for surface access to the Crown land where Amoco proposed to drill its well. After a site visit by field staff from Fish and Wildlife and Public Lands, the government granted access on condition that no drilling take place between January and April, the time when wintering elk would visit the area. Government field staff do not regard it as their role to allow or disallow drilling in areas such as the Whaleback that are not specifically protected by government policy. Instead, they see their role as trying to help an applicant modify its development plans to mitigate environmental impacts.

The government's mineral leasing process shifted the question of whether Amoco could drill the Whaleback onto the ERCB, and the ERCB turned Amoco down because it detected a conflict with potential government policy. Critics charge that the government should detect such conflicts much earlier in the process.

"It's a government screw-up," says Cliff Wallis, president of the AWA. "Amoco never should have got those mineral rights in the first place. The government has known the Whaleback has been one of our top 10 priorities for protection as a wildlands recreation area since the early '80s."

Wallis believes the public should have input to the Crown Minerals Disposition Review Committee the way it does to the ERCB and the NRCB. "The ERCB provides us with a chance for input on surface rights but the government will not on mineral rights. We've asked for it but the government is intransigent in its refusal."

In its decision, the ERCB says it expects the government to implement the Special Places 2000 program "in a timely fashion," but so far the government has announced no date for cabinet review of the policy now in the hands of the minister for Alberta Environmental Protection. Many environmentalists, including Wendy Francis, say they think the ERCB's acknowledgement of the program will help move it successfully through cabinet. Sources inside the department fear that all the attention focused on the program may have the opposite effect.

As for the Whaleback and its protection as a special place, no one makes any promises.

Never Give Up

Amoco isn't prepared to give up just yet. It's as though they had bought a car and then gone to get their driver's licence but failed the test. What would you do? Wait a couple of days and try it again– that car sitting in the driveway is serious motivation.

And the ERCB certainly didn't close the door. Here's how it concluded its decision report: "The Board acknowledges that the natural gas prospects in the Whaleback area may be promising and evaluation of the prospect may eventually be in the long-term public interest of Alberta... The Board is prepared to consider a new application upon clarification of the land-use status of the area and subject to the submission of further evidence on the issues identified in this report."

Unless the provincial government sets out a policy clearly protecting the Whaleback from mineral exploration and development, it's as though the applicant just failed parallel parking and will be back once he's practised up.

Less than two weeks after the ERCB's decision, Amoco was back in the Whaleback, talking to the local folk about the alternatives. Their reception wasn't all that warm but they're motivated. "We're reviewing how we'll enhance our program," says Vince Rodych. "We're looking at access management and well location among other things. We will likely reapply."

ON THE WHALEBACK'S ECOLOGICAL SIGNIFICANCE

In November 1995, the Heritage Protection and Education Branch of Alberta Environmental Protection responded to the ERCB decision. The government report, *Alberta's Montane Subregion, Special Places 2000 and the Significance of the Whaleback Montane*, spoke to the concerns the ERCB had about whether or not the Whaleback would be protected from industrialization. The ERCB had speculated in its decision that the Whaleback would be a "prime candidate" for protection in what was then a very nebulous initiative – the province's Special Places 2000 policy. The ERCB concluded:

> "The board believes that the Special Places 2000 process is the logical forum in which to debate the overall public value of the Whaleback from the provincial perspective. In the absence of such an evaluation having been performed at this time, the board does not believe it would be in the public interest for it to approve an application for energy development that may, in turn, significantly compromise a scarce or unique combination of ecological values."

The following excerpt from the government report confirms the ecological significance of the Whaleback and argues that extending protected status to the Whaleback would make an important contribution to the Special Places initiative.

I.U.

"3.2 The Significance of the Whaleback

Maps... show that there are few blocks of intact montane landscapes remaining. Prominent among those remnants is the Whaleback area... It is the largest block, being almost twice as large as any other remaining block, with only the South Ghost... and the North Porcupine Hills... coming anywhere near the size of the Whaleback. Both those latter sites, however, lack the wilderness character of the Whaleback as a result of moderate to high numbers of OHV (Editor's note: Off Highway Vehicles) trails and other impacts. In addition to its size and wilderness character, the Whaleback remains connected to adjacent areas. That is not to say that the Whaleback area is immune from a certain amount of OHV use or other activities. Those activities do occur here.

The Whaleback area, with its approximate 22,400 ha of relatively intact and high quality montane landscape contains a number of outstanding natural values. This area represents the classic montane by containing the characteristic plants, animals and natural communities associated with the Montane Subregion (Bradley et al. 1977).

The natural values of the Whaleback area were identified from a number of sources including herbarium data, rare plant data from the Natural Heritage Information Centre of Alberta Parks, information on file with the Natural Heritage Protection and Education Branch of Alberta Parks, maps and biophysical reports (Brown et al. 1986a, 1986b; Cottonwood Consultants 1983, 1987a; Downing and Karpuk 1994; Strong 1979; Wallis 1980, 1994).

The outstanding natural values recognized for the Whaleback include the following:

- essentially the last remaining "wilderness" montane (most extensive and least disturbed East Slope Rocky Mountain Montane in Canada);
- spectacular scenery;
- some of the most extensive limber pine and Douglas-fir stands in Alberta; and
- the most productive wildlife area in foothills region, featuring the following:

  - diversity of breeding birds including Golden Eagle;
  - carnivores including grizzly, wolf and cougar;
  - key mule deer habitat;
  - critical moose winter range;
  - provincially significant and critical elk winter range;
  - elk calving and summer range;
  - one of the highest densities of cougar populations in Alberta;
  - superior trout stream with mountain whitefish and cutthroat and bull trout;
  - overwintering habitat for bull trout; and
  - the Oldman River below the Gap is recognized as a "Class 1" trout stream.
- representative montane communities:
  - riparian woodlands
  - rich aspen woodlands
  - diverse grasslands
  - diverse forb meadows
  - limber pine stands
  - Douglas-fir stands
  - aquatic communities
  - Balsam poplar stands
  - Engelmann spruce stands

In addition to these outstanding values, a number of special features are found in the Whaleback... These special features include:

Geological/geomorphological features:
  - large glacial erratics
  - well-preserved examples of abandoned stream terraces
  - fossils
  - geological sections along the Oldman River
  - areas of pitted outwash
  - outcrops and rugged ridges, including the Whaleback Ridge

Rare plant species:
  - 22 rare species have been documented in the area

Rare or unusual animal species:
  - golden eagle and prairie falcon nesting sites
  - pika colonies (outside their usual range)

Rare or unusual plant species:

- saline seepate areas
- springs
- lush forb meadows
- extensive low willow/swamp birch communities
(relatively rare in the montane)

The Whaleback is a significant area for wildlife, largely because it remains connected to a larger ecosystem and the nature of the landscape facilitates wildlife movement... The Whaleback "Study Area"... which encompasses portions of the Subalpine Subregion as well as important portions of the Montane Subregion, provides a core of essential habitats to a number of large mammals that range over a much larger area.

The Whaleback is considered a provincially significant elk winter range... Although some of the animals are part of a resident herd that also summers in the Whaleback, some winter in the Whaleback and summer to the west of the Livingstone Range – some as far west as into British Columbia.

Although the patterns of use by elk have not been adequately studied, there appear to be two main movement corridors: one along Deep Creek, the other along White Creek... However, animals are not restricted to these corridors for their travel. They can move throughout the area and, depending on weather and snow depth, are not constrained by topography. Some movement likely occurs along the Oldman River, but the presence of roads and the narrowness and open corridor at the Gap may inhibit use.

Some ridges such as Chaffen, Black Mountain, East Ridge and the ridge west of the Whaleback probably provide secondary winter range, particularly in low-snow years. The entire lower Bob Creek valley and adjacent ridges are critical elk winter habitat as well as important calving areas and summer habitat. Additional calving areas may be present in the upper White Creek valley, but this requires verification. There is a further need to identify elk rutting areas and mineral licks.

Although elk are the best-studied of the large animals which use the Whaleback and nearby areas, there remain a number of questions about their use habits, movement patterns and about the location of critical habitats. Even less is known about the other large mammals in the area. The presence of cougar and the continued use of the area by grizzly bears are likely due to the roadless nature of the area. It is important for these species, as well as for elk, that linkages are maintained between the Whaleback and adjacent lands, particularly lands to the west.

The importance of the area, particularly to large mammals, is strongly related to maintaining both the montane and the adjacent block of subalpine to the west as a single, intact and functioning landscape.

Protection of the natural values of the Whaleback area would make an important contribution toward the completion of the Natural History Theme targets under the Special Places 2000 Program..."[8]

HARPOONING THE WHALE

The text on the current issues webpage of the Calgary/Banff chapter of the CPAWS urges readers to:

> Be optimistic! Emphasize the successes that are being achieved in protecting Canada's Endangered Spaces (Tatshenshini!– Wind Valley!– Whaleback!) in order to remain positive about the possibility of progress.

Writing in July 1998 it looks like the Whaleback will soon be dropped from this short list of successes. Special Places, the process where the ERCB felt the Whaleback would be "a prime candidate" for protection, seems instead to be poised to open the Whaleback to petroleum exploration and development. This outcome seems likely despite the government study noted above, despite the ERCB's invitation to extend protected status to the Whaleback, and despite the willingness of a petroleum industry stung by the Amoco decision to set aside areas which would be protected totally from industrial use.

How could Special Places transform the Whaleback from an issue CPAWS brags about to one the organization may be quick to condemn? Two characteristics of the Special Places 2000 process – the importance of local committees and the government's promise to honour existing land use commitments – push its decision-making in a direction likely to privilege economic development over the preservation of natural landscapes and change. These two culprits are highlighted below in James Tweedie's explanation for why CPAWS ultimately resigned from the Whaleback Local Committee in December 1997.

Ty Lund, the Minister of Environmental Protection, underlined the important role he wanted local committees to play in the fall of 1997: "Albertans in the communities near these candidate sites are best suited to recommend boundary options and appropriate land-use activities for each candidate site and to develop site-specific management principles."[9] What made local committee members, such as municipal politicians or off highway vehicle users, better suited than environmental scientists to assess the impacts of various land-use activities on the Whaleback was never explained.

Lund's department also made it clear on numerous occasions that existing development commitments will not be sacrificed on the Special Places altar. "Under the Special Places initiative," his department promised, "the government will continue to honour existing commitments to the oil and gas industry when new protected areas are identified and established... The selection of new sites will consider both ecological values and impacts on the natural gas and petroleum industry."

The Minister's high regard for the importance of local committees and existing commitments explains the wide open terms of reference the Whaleback Local Committee received. When the committee was established it was asked to recommend to the Minister, "permitted activities (uses) and guidelines for ongoing activities within the Special Place(s), including a rationale for those uses consistent with meeting the preservation goal."

The portion of the Committee's draft report dealing with oil and gas development– an excerpt which favours opening the door to development– follows Tweedie's commentary. I.U.

THE WHALEBACK–WHY WE LEFT THE TALKS[10]

-by James Tweedie

In September 1994, the Energy Resources Conservation Board (now Alberta Energy and Utilities Board) issued a Decision Report on Amoco's application for exploratory drilling in the Whaleback. It had this to say about the area: "The Board believes that the Whaleback area represents a truly unique and valuable Alberta ecosystem with extremely high recreational, aesthetic, and wildlife values. It accepts the position of some interveners that the area is a primary candidate for protection under the provincial Special Places 2000 program. A significant component of that value lies in the relatively large and contiguous nature of the Whaleback Ridge ecosystem and the very limited disturbance that has occurred."

It was to see the fulfillment of that protection goal that CPAWS, with Friends of the Whaleback, agreed to participate in a Local Committee established in February of 1997 to review the nomination of the Whaleback to Special Places 2000.

A local rancher was nominated to represent the Federation of Alberta Naturalists. Of the remaining 11 seats, half were taken up by local ranchers, the remainder going to representatives from forestry, oil and gas, pipelines and utilities, off highway vehicle users and local outfitters/guides/ trappers. Native representation was invited but declined.

Unfortunately, this process failed. Rather than use the opportunity of this process to develop a shared vision for the Whaleback, the committee spent most of its time hearing from those with the greatest concern to protect their vested interests in the area, and from the administrative departments of the Provincial government responsible for managing the area for multiple use. The Government's decree that "all existing dispositions be honoured", without any discussion of possible ways to accommodate the needs of the "protection goals", meant the committee's work was fundamentally flawed from the outset. Deliberations were diverted from the "science-based" approach to land-use planning touted in the Special Places policy to a continuation of the multiple-use approach taken by the existing IRP.

The ranching community, which dominates the committee and which owes its tenure of the land to the Department of Forestry, appeared unwilling to risk making any decisions that might alienate that department. The Local Committee therefore chose to spell out its own meaning of "protection" for this area: "protection" for all current activities, as in the case of the present extensive grazing regime, and the endorsement of industrial and forestry activities that may potentially arise in the future. No serious attempt was made to explore alternatives to the existing allocation of future timber quotas. The oil and gas industry representative pushed for the endorsement of his industry's agenda in the area by this committee, despite earlier commitments by Amoco and CAPP to live with the decisions that would fall out from the committee's deliberations. The underlying rationale for all decisions appeared to be that since the ranching community owed its livelihood to the commercial exploitation of grazing it could not oppose any other commercial activities.

After nine months of exasperation at being unable to keep the committee focused on the "protection goals" of the Special Places policy, stated in the original mandate for the committee, CPAWS and the Friends of the Whaleback have resigned from the committee. The committee is expected to submit its recommendations to the Minister in the spring. Sadly, they will probably include recommendations regarding oil and gas exploration and development and commercial logging that have no place in any "protected area".

WHALEBACK LOCAL COMMITTEE DRAFT RECOMMENDATIONS [11]

"Oil and Gas Resource Development

18. The Local Committee agreed to the following statement of intent for oil and gas activities:
    - The Local Committee agrees that commercial oil and gas exploration and production operations are acceptable activities.
    - A local advisory board should be established and consulted at every phase of an operation and continually participate in discussions of operations issues, safety issues and current or future planning.
    - Any oil and gas activities would require a common operator approach, such as an operating council, for the area as a whole and address the following:
        - Consultation with the local advisory board;
        - Operation and coordination of all oil and gas activity in the area;
        - Environmental and reclamation planning and implementation;
        - Oil and gas activities will be done in a manner which recognizes the environmentally sensitive nature of the area and maintains or improves the ecological health of the area over the long term;
        - Certain unique terrain may be considered off limits to land disturbance i.e. Whaleback Ridge...
        - There should be no long term disruption of wildlife or wildlife corridors;
        - Plans should ensure the protection of endangered species;
        - Motorized access should not be increased over the long term and should be managed over the term of the operation;
        - Reclamation should be using native seed and plant stocks;
        - There should be no sour gas processing plants situated within the area;
        - Visual impact of buildings, well heads and other operations equipment should be mitigated.

The Local Committee is in agreement in their concern about the methods used to extract non-renewable resources. The list of guidelines provided in the recommendation is not intended to be all inclusive. The Local Committee expects operators in the recommended designated area to cooperate, to minimize impacts using all available means, and to be accountable to the Local Management Committee in all of its operations.

The Local Committee approved this recommendation by a vote of 9-4. The Chair, both representatives of environmental interests, and the representative of recreational fish and game users voted against this recommendation."

KANANASKIS COUNTRY

Kananaskis Country is a 4,156 square kilometre recreational area less than an hour's drive southwest of Calgary. Created by the provincial government in 1977, K-Country now sees more than 2.4 million recreational users every year– the majority from the metropolitan Calgary area. Although Kananaskis is a recreational area, the area has been managed by the multiple use concept. Recreational users share this space with loggers, ranchers, and the oil and gas industry. Currently, the future of recreational development is being studied by the provincial government. Until the government's policy review is completed (expected completion: Fall 1998) there is a moratorium on new recreational development in Kananaskis.

I.U.

"Without them, many of the conservation gains we've made wouldn't have been possible. They have a huge impact on what we do."

- Larry Simpson, Nature Conservancy of Canada,
on the petroleum industry and the efforts of the NCC to conserve land.[12]

QUIRK CREEK[13]

- by The Alberta Wilderness Association

In 1969 the Alberta Wilderness Association first considered the area at the headwaters of the Sheep and Elbow Rivers for Wildland protection. In 1972 the AWA, during the Eastern Slopes hearings, suggested the creation of the Elbow-Sheep Wildland Park to encompass both mountains and foothills in a protected core reserve. The hearing board accepted the concept of the Elbow-Sheep Wildland Park, but the government failed to act on that recommendation for over 20 years. In 1996 the AWA Wildland proposal was expanded, based on new scientific data on cougars, moose, wolves, grizzly bears and bull trout, to include lands found within Kananaskis Country as far east as the Forest Reserve Boundary, north to McLean Creek Off-Road Vehicle area, and south to Highway 541.

Roughly half of this area was protected in 1996 when the Minister of Environmental Protection Ty Lund created the Elbow-Sheep Wildland Park, under the auspices of Special Places 2000. This protected area, however, failed to include any of the sensitive foothills natural regions of Kananaskis Country. Instead it protected only high elevation alpine environments, and barren alpine rock. At that time the Minister, in a letter to the AWA, stated that boundary considerations would be made under the planning process for the Elbow-

Sheep. The AWA has since asked that their revised Elbow-Sheep Wildland boundary be adopted as an amendment to the area management plan.

In June of 1993 Husky Oil of Calgary Alberta acquired a 32 section P&NG License (oil and gas lease) for $8.2 million. This lease is a five-year lease to explore for oil and gas in the heart of the AWA's long standing Elbow-Sheep Proposal. The lease expires in June of 1998.

During the summer of 1994 Husky Oil shot 77 kilometers of seismic though (sic) out the lease, and as a result determined that a possible play was located above the Turner Valley Formation in the area of Three Point Creek, between Volcano Ridge and Forget-me-not Ridge. Husky suggests that the play is located in the Mississippian Formation, 1100 meters below the surface, and that it could contain either oil or gas. Husky estimated that there were 40-50 million barrels present in this play, with 6-15 million barrels recoverable.

Between the summer of 1994 and spring of 1997, Husky Oil conducted extensive biological surveys of the region, with the intent of using the information as background in an application to drill a total of 7 wells in the Three-Point Creek area. These wells would be drilled straight down, and would be accompanied by 2 water injection wells. In order to access this previously undisturbed site, 20 kilometers of road along route 66 would be upgraded (through the McLean Creek Off Highway Vehicle area), and 5 kilometers of new road would be built along the Wildhorse Trail and Volcano Creek. A new 20-kilometer long pipeline would be constructed along the Hog Back Ridge and Three Point Creek to tie the operation into the Quirk Creek Plant, located just outside the Forest Reserve Boundary.

During the course of Husky's biological survey, it was determined that the area around Three Point Creek was of great significance environmentally, and the company made the decision to ask the government of Alberta to accept the return of the 8000 hectare Quirk Creek lease, and reserve it from re-sale for oil and gas exploration. Husky Oil made this decision based on its own studies of the area, in which they concluded that "habitat loss or alteration, increased access, the impacts of roads and traffic and other disturbances" would severely impact "elk, mule deer, bighorn sheep, grizzly bears and cougars."

Husky clearly recognizes the importance of the foothills of Kananaskis Country as wildlife habitat, and as the source of Calgary's water supply. They also recognize that the public is against further destruction of Kananaskis Country, a place many Calgarians and Albertans think of their back yard.

In July of 1997 Husky Oil executives met with Minister of Energy Steve West, and Minister of Environmental Protection Ty Lund. At that time, Husky Oil presented the Ministers with the findings of their seismic research and their biological survey, and concluded the talk by asking the government to accept the return of the lease in exchange that (1) it not be re-leased for oil and gas exploration and (2) that Husky Oil be issued a credit on their next purchases of an oil and gas lease anywhere in the province of Alberta.

In a one-paragraph letter from Minister of Energy Steve West the government denied Husky Oil's appeal to return the Quirk Creek lease. The letter was sent sometime in August or September of 1997.

Section 8(1) of the Mines and Minerals Act states that

> "The Minister may... (c) accept the surrender of, cancel or refuse to renew an agreement as to all or part of a location when the Minister is of the opinion that any or any further exploration for or development of the mineral to which the agreement relates within that location or part of it is not in the public interest, subject to the payment of compensation determined in accordance with the regulations for the lessee's interest under the agreement."

Husky Oil now has four options, in the opinion of the Alberta Wilderness Association:

- Continue to lobby the government to accept the surrender of the Quirk Creek lease. Husky Oil has indicated to the AWA that it will ask the Premier to intervene in this matter,
- Drill a well in the Three Point Creek area to determine whether or not there is a play above the Turner Valley Formation, or
- Ask for an extension of the Quirk Creek lease, but delay the drilling of a well at Three-Point Creek.

Refuse to drill a well, allow the lease to expire, and offer as advice to all other oil and gas companies with interests in the foothills of Kananaskis Country that the Quirk Creek lease contains significant wildlife resources, and that the environmentally sensitivity is considerable (sic).

The AWA would prefer that Husky Oil exercised options 1 and/or 3. If Husky Oil decides to drill a test well in the Three Point Creek area, the AWA will likely intervene at and (sic) Energy Utilities Board Hearing, and will appeal to Albertans to protect the foothills of Kananaskis Country.

The Alberta Wilderness Association applauds the efforts of Husky Oil to protect this valuable section of Kananaskis Country, and hopes that the government will take its lead from industry in this extraordinary case.

Editor's postscript: The Alberta government allowed Husky to extend its Quirk Creek lease until June 2001 (the lease had been set to expire at the end of June 1998). Stephen Legault of the Alberta Wilderness Association called this decision "a stay of execution for Three Point Creek. It gives us some hope. We have an opportunity now to continue our work to protect the foothills of Kananaskis Country without the immediate threat of drilling in this sensitive and wild region."

CHAPTER 4

GRIZZLIES, COWS N' GUNS

Oh how I love to gaze at the mountains, How I love
to swim in the streams,
In those crystal pure, icy cold fountains, those rivers
I've travelled and seen.
Those pictures which God's hand has painted,
crowned with the setting sun's beams
And steeped in the nectar of nature, are in this life
like a beautiful dream.

- from "When the Saw Flies Mate in Summer"
by Henry Stelfox[1]

— INTRODUCTION —

Indicator species, umbrella species – the outdoors magazines I devoured as a kid never used these phrases to describe grizzly bears. In the "hook and bullet" press where I first encountered grizzlies, stories about Ursus arctos horribilis stressed the horribilis – the fierceness, the incredible strength of this species of the bear family. The grizzly symbolized the wildness, the power, and the ferocity of nature. The grizzly took a place as one of our most feared non-human antagonists in a "man versus environment" narrative, a narrative where the successful hunter confirmed our dominion over nature.

Today, although the grizzly still conjures up the awe and the fearsome imagery of thirty or forty years ago, much is different. Ecologists label the grizzly as perhaps the most important indicator or umbrella species for the ecological integrity of the eastern slopes of the Canadian Rockies. The health of the grizzly bear population is regarded as a key to maintaining healthy ecosystems. "If grizzly bears survive in an area," notes Dr. Stephen Herrero, "then you can fairly well be assured that the area has a fair degree of ecological integrity which allows us also to think of grizzly bears as an umbrella species. Maintaining grizzly bears will encourage terrestrial ecosystem integrity."[2] Look after the habitat needs of grizzlies and you look after the habitat needs of most of the mammals found in the Rockies and Foothills.

Unfortunately for the grizzly, the habitat it needs also is sought by virtually every human activity discussed in this collection. City dwellers, resort developers, loggers, miners, ranchers, and oil company executives all want slices of the territories through which grizzlies must roam. The material in this section considers two of the activities – ranching and hunting – which have been accused of contributing to the decline of Alberta's grizzly bear population, a population the provincial government has declared "at risk." The articles, by the award winning writers Bruce Masterman and Kevin Van Tighem, stress the importance of establishing secure bear habitat for the future of this wilderness icon. Without a government commitment to establish the territorial integrity the grizzly needs, the bear's future is grim. I.U.

POLL HAVEN: GRIZZLIES TAKE A STAND [3]

- by Bruce Masterman

Since the days when European man first ventured west, man and the Grizzly Bear have shared a rather tenuous relationship. In the man-imposed design of the west, the great bear has been allotted but a few pockets of wilderness in which to live undisturbed. And more often than not Grizzlies have found themselves at odds with people. From a time when Grizzlies thrived over a vast landscape, ranging from Mexico to Alaska, their numbers have rapidly dwindled in the face of logging, recreation development, oil and gas projects, poaching and agriculture. The Grizzly has been pushed to the limit in North America – and then some.

But in a small corner of southwestern Alberta, the Grizzly Bear is pushing back. And ranchers have reacted with outrage.

The Poll Haven community pasture covers 4,096 hectares (10,240 acres) of rugged poplar and Jack Pine forest abutting Waterton Lakes National Park to the west, Montana to the south and Montana's Glacier National Park to the southwest. Less than 40 years ago, Poll Haven was part of Waterton Park. Now it's provincial Crown land, used mainly by 40 local ranchers who hold exclusive rights to graze their cattle in the area for four months each spring and summer.

But the cattle are not alone at Poll Haven. Grizzlies from Waterton and

Glacier parks wander in and out of the area on a seasonal basis. Usually the bears pass harmlessly through the pasture, minding their own business as they forage for succulent native vegetation that makes up the bulk of their diet. Every so often, though, the Grizzlies get a taste for beef instead of salad. That's when they take advantage of all that hamburger-on-the-hoof that's so readily available in the remote pasture. The ranchers blame Grizzlies for killing 100 head of cattle on and around the community pasture since 1983. Many of the cattlemen are calling for blood: the Grizzlies'.

Jan Allen is no stranger to Grizzly or Black Bears. A veteran Alberta Fish and Wildlife officer, the 49-year-old Allen is widely recognized as an expert handler of problem bruins. He's been doing it for 24 years. "I probably owe this to bears," Allen grins as he runs a hand through his thinning, silver-grey hair. Although he prefers to keep a low profile on the job, Allen has found himself thrust front and centre into the Poll Haven controversy. Allen is responsible for catching Grizzlies suspected of killing cattle in southern Alberta, including Poll Haven. He walks a public relations tightrope, acting as a referee-ambassador to ease the ranchers' bitter feelings toward Grizzlies while also trying to deal with problem bears safely and swiftly. He leaves the politics of the issue to the politicians. Allen fears that if he doesn't respond to the ranchers' complaints, some of them would handle the problem arbitrarily – with guns. Although one rancher suggests such action would be justified there's no evidence to suggest it's happening, although one rancher opined, "I think they have just cause for it if they are."

A quick capture of a cattle-killing Grizzly is crucial. "Every day the bear isn't caught the situation becomes more dangerous, more volatile," Allen says. "We've all heard stories about people being mauled after surprising a Grizzly on a kill." During the summer, the telephone rings often in Allen's home and in his government office in Pincher Creek. Often it's a rancher reporting a dead or injured cow. Such a call never fails to put Allen's adrenaline into overdrive. Grizzlies do that to a person, hardened veteran and novice alike.

One rainy morning in mid-July of 1987, a call came from rancher Lee Nelson. He reported that a Grizzly had stepped into a spring-loaded leg snare the night before. Allen had set the snare a few days previous after Nelson found one of his cows with deep gashes in its flank suggestive of bear claws. The cow's calf was missing. Circumstantial evidence suggested the cow had been injured while unsuccessfully defending its calf. After Allen hung up the phone, he quickly contacted a team of fellow Fish and Wildlife officers from other southern Alberta offices.

Allen, a leader of one of the division's four special bear response teams, opened a padlocked trunk and issued each team member a .357 Magnum revolver. Bear response teams are the only Alberta Fish and Wildlife officers permitted to carry sidearms. Ordinarily, officers carry 12-gauge shotguns and 30-06-calibre rifles in their vehicles. Allen says sidearms are more useful because they are worn in a holster, leaving an officer's hands free to set a snare or check a cattle kill, but still handy enough to use in a crunch. An unarmed person is no match for an angry Grizzly Bear capable of snapping the neck of a full-grown, 409-kilogram (900 pound) cow.

"I hope I never have to use a sidearm," Allen says. "It is positively the last resort."

On that summer morning, tension was thick inside the three green-and-white Fish and Wildlife trucks as they slowly snaked along a muddy trail, threading through thick stands of pine and poplar. As the cavalcade entered a small clearing, it was greeted by an awesome scene: it looked as though a small tornado had struck.

Underbrush and willows had been uprooted and tossed about in the forest on the other side of the clearing. Poplars several inches thick had been snapped like toothpicks; thicker trees had long, deep gashes and were gnawed halfway through. The source of the devastation was lying partially hidden in a shallow crater gouged out of the forest floor. A shaggy, chocolate brown Grizzly of about 227 kilograms (500 pounds) watched us quietly, exhausted from a long night spent struggling to free its right front leg from a heavy steel cable anchored to a stout tree. The bear's eyes blazed defiantly, but at the same time they seemed to reflect some plotting as to how it could make good an escape. While the other officers trained their slug-loaded shotguns and high-powered rifles on the trussed Grizzly, Allen prepared the tranquilizer gun. This bear was in a relatively easy-to-reach spot. Allen knows too well the feeling of having the hair on the back of his neck stand straight up while he's crawling through jungle-like vegetation to check a snare hidden deep in bush where a Grizzly is suspected of being. The added uncertainty of not knowing whether a Grizzly is indeed caught – or how secure it might be – makes for some tense moments. But Allen and his officers have never been injured during a capture.

Suddenly the Grizzly stood. It suspiciously eyed the officers poised 60 metres, just several bounds, away. Index fingers tightened against triggers; nerves were taut. The bear charged. Right now, without warning. All eyes were riveted on the cable, which suddenly appeared as thin as thread. The rage-powered bear stretched the cable to the limit, releasing an angry bellow.

The observers tensed, everyone thinking the same questions but having no time or capability to vocalize them. Would the cable hold or had the bolts securing it to the tree somehow loosened? Had the anchor tree been chewed through during the night? The answers came mercifully fast as the bear was snapped back by the taut cable. Defeated, it retreated into the trees, while uttering a low growl that rumbled across the clearing. The collective sigh of relief among the sideline observers was silent, but very perceptible.

Backed by his armed officers, Allen carefully moved in closer and fired the dart into the bear's rump. It angrily whirled to face the source of the sudden sharp sting, but the demobilization drug started to take effect almost immediately. Within minutes the big bear was sprawled in sleep. The officers moved in. An examination shone no light on the bear's sudden appetite for beef, apart from its easy proximity.

The Grizzly was a healthy male with strong teeth and long scimitar-like claws that could help bring down almost any prey it chose. Now the bear had become a suspected cattle killer. Its wanderlust and appetite had conspired to bring about the Grizzly's humiliating capture.

The Grizzly had earned himself an unwanted free trip in a steel cage on wheels, courtesy of the province of Alberta. His destination was a strange place in the mountains hundreds of miles from home, far from ranchers and temptations of the bovine kind. By trying to survive in a world too-small, the Grizzly had committed a grievous crime against man. The bear's sentence was banishment, but it could have been much worse.

Not long ago, the Grizzly would have been summarily executed.

Ranchers say there were few problems with Grizzlies in the area 20 years ago when they were allowed to either shoot the bears or hire professional hunters to trap and kill them. But the Grizzly has been given some measure of protection in Alberta – as it has in the few other North American jurisdictions where it still roams, and frontier justice is no longer acceptable. But some ranchers have issued a call to arms. They believe they should be permitted to use firearms in this fierce tug-of-war with the Grizzlies over their livestock.

Rancher Nelson, who runs about 75 head of beef cattle on a half-section of land abutting the community pasture, says, "A person has to protect his property. The rancher ought to have the right to protect his stock."

Alberta's environmentalists and the provincial government disagree. "We're at the turning point where we have to decide for the bears or more cows," says Brian Horejsi, a Calgary wildlife biologist who has conducted extensive Grizzly research in northern Alberta. Former vice-president of the 2,500-strong Alberta Wilderness Association, Horejsi is leading the fight to save the Grizzlies in southern Alberta.

Horejsi is particularly critical of the provincial government's response to the problem. The Alberta Fish and Wildlife Division attempts to catch any Grizzlies suspected of killing or injuring cattle. Duane Radford, the division's Lethbridge-based southern regional director, says, "Once they start killing cattle, they don't stop until they're (live) trapped." From 1985 to 1987, 18 Grizzlies were captured in Fish and Wildlife snares. In the summer of 1987 alone, six male bears were caught. That none were captured in 1988 is officially attributed to a rich berry crop; critics say it's because few bears are left in the area.

No complaints were registered again in 1989, leading Radford to conclude that the Grizzly-cow conflict "is unlikely to happen again in the foreseeable future." He suspects past problems were caused by a large number of older, debilitated bears who'd turned to beef for an easy meal.

Some bears were cattle killers, simply killing cows without feeding on them, Radford says.

Horejsi outright dismisses this logic as "a bunch of crap." He said a 15-year-old is in its prime and isn't likely to start killing cows unless it is feeling other pressures on its territory.

The fact that no complaints surfaced in the past two seasons doesn't surprise Horejsi. Since 1979, he says, 77 Grizzlies have been either killed or relocated from the Poll Haven area. Some of them were taken by hunters. "I don't expect there are many Grizzly Bears left there now." The province's Grizzly Bear management plan, still awaiting ministerial approval, basically downgrades the Poll Haven district from prime to secondary Grizzly habitat. Horejsi blames man's encroachment for the change in classification.

"It's an assault on an ecosystem that we haven't seen before," he says. "And Grizzlies are taking the fall."

Of the bears caught in 1987, half were relocated to remote regions of the Rocky Mountains in west-central Alberta, including a massive 331-kilogram (728-pound) boar. The other three were killed. Horejsi resists any plan to move Grizzlies out of their natural habitat to a strange area, and he's incensed that some of the bears have been killed. One boar was judged too old to survive relocation; another suffered a broken leg while struggling in the cable snare; the third was weakened from a radio-collar of undetermined origin that had become too tight, preventing the bear from feeding properly.

To Horejsi, it makes no sense to kill or move every bear that crowds man a little. He advocates total elimination of grazing on Poll Haven and the government purchase of about 25,600 hectares (64,000 acres) of private land around Waterton Park – at an estimated cost of $50 million – to create a refuge for the Grizzly. Similar action on a smaller scale was taken in Montana when the U.S. Nature Conservancy, a private, non-profit organization, bought 8,800 hectares (19,360 acres) of land just south of Glacier Park and the Blackfoot Indian reservation.

Horejsi's proposal has met with a cool to tepid response from ranchers, many of whom have roots in the area dating back more than 50 years. Some say they're not about to be forced out by a few Grizzly Bears. Some are prepared to negotiate.

The provincial government is drafting up an integrated resource plan for an 80-square-kilometre area in and around the Poll Haven pasture. It calls for continued use of forage and timber resources and the protection of ecological, fisheries, historical, water and wildlife resources. Extensive recreational use is featured.

Unsurprisingly, Horejsi disagrees, protesting that protection of Grizzlies should have top priority.

"When you have a piece of land that was largely Grizzly Bear habitat in past years and you give a little piece to everyone, or a large slice to cattlemen, the Grizzly is hurt," says Horejsi.

One warm evening in August 1987, Horejsi and a few other friends of the Grizzly found themselves in unfriendly territory. About 100 people, most of them range-hardened ranchers, packed a smoky hall in the provincial building in Cardston, a small town east of Poll Haven. They were there to discuss the Grizzly-cattle conflict. One rancher set the tone by stating, "We've got a severe problem, and it's costing us a lot of bucks."

Broyce Jacobs, reeve of the Municipal District of Cardston and himself a rancher, expressed the cattlemen's frustration that they're expected to incur losses from bears and they can't legally protect their cattle. But, he added, none of the ranchers "want to drive the Grizzly Bear to extinction." Nonetheless, the mood of the meeting remained anti-bear.

Horejsi evoked a round of catcalls from the gathering when he suggested no action should be taken against a cattle-killing bear until it's struck twice. He urged ranchers to consider Grizzlies a valuable resource worth preserving. Decisions on their future must be made with regard to the interests of society as a whole, not just the ranching community, he said.

Another Grizzly supporter said the ranchers were inviting the wrath of conservationists across Canada by making the issue "Alberta's seal hunt of the 80s."

Not all ranchers resent the Grizzly. Third-generation cattleman Mike Dawson, 48, concedes that locally, his is a voice in the wilderness. He maintains ranchers shouldn't have the right to shoot Grizzlies. Instead, he says, they should accept that they live in bear country and will lose the occasional cow to a Grizzly.

"If you go to Rome, you're liable to meet a Roman," he says. In a letter to the Calgary Herald, Dawson's wife, Bev, wrote: "We ranchers put our cattle practically on the bears' table and then complain if a bear gets one or two... I feel it's a price I must pay for invading the bear's domain."

A major sore point among ranchers is that they don't receive full compensation for cattle killed by Grizzlies. The government pays 80 percent of the value of any cow or calf confirmed killed by a Grizzly. Fifty percent is paid for a probable Grizzly kill while a missing animal nets 30 percent.[4]

"If a cow is dead, it's 100 percent dead, not 80 percent dead," says rancher Nelson.

Because of the community pasture's proximity to Montana, the Poll Haven issue has crossed the international border. It's believed some of the beef-eating bears being moved or destroyed by the Alberta government live in Glacier National Park at some time of the year.

"The bear knows no limit between the boundaries," says Loren Kreck, president of the Flathead chapter of the Montana Wilderness Association. Grizzly Bears, he suggests, should be treated as a valuable international resource to be shared equally by Albertans and Montanans.

Kreck is quick to point out the Grizzly was wiped out in New Mexico, Arizona, Colorado and California mainly because it conflicted with ranchers. It's estimated fewer than 35,000 Grizzlies remain in North America. The most concentrated populations are in Alaska, the Northwest Territories and the Yukon. Alberta has about 600 Grizzlies outside the national parks – compared with 50,000 black bears – while Montana, Wyoming and Idaho together have fewer than 1,000. Officially, the bear is listed as a threatened species in those U.S. states.

History has proven ranchers have been largely responsible for the extirpation of the Grizzly Bear over much of its original range in the United States. Cattlemen and sheepmen demanded bounties on Grizzlies and the bears were shot, poisoned or trapped. It's commonly believed many Grizzlies may be wrongly blamed for killing livestock only because they are found feeding on the carcass of a cow they discovered after it had died from other causes.

Gary Gregory, resource manager of Glacier Park, is upset by the number of Grizzlies that have been removed from Poll Haven. He's particularly concerned that loss of several adult male bears could jeopardize the Grizzly's future in northern Montana. It's estimated more than 200 Grizzlies frequent Glacier Park at any one time.

"Grizzly Bears are a powerful force in the ecosystem and that's why it's so important they be protected," says Gregory. The issue has also sparked concern in neighboring Waterton Park, which has an estimated Grizzly population of 25.

Bernie Lieff, who served as Waterton's superintendent for seven years until 1988, agrees it's important to keep Grizzlies in the district rather than continuing to trap, relocate or destroy them. The Grizzly, Lieff believes, will inevitably get into trouble with cattle grazing on the park's borders and with oil, gas and recreational developments crowding the park's edges. All this activity restricts the bears' natural movements and heightens the chance of conflicts. Lieff challenges the Alberta government to work with Montana to develop an overall management plan that protects Grizzlies whether they are in Alberta or south of the border.

The Poll Haven issue, he says, is a long-term problem we're not dealing with in a long-term manner. He adds that it's impossible to confine Waterton's Grizzlies within the park's 525 square-kilometre area.

"The only way you're not going to have bears leaving the park is not to have any bears in the park."

THE GRIZZLY HUNTING PLACEBO[5]

- by Kevin Van Tighem

It was probably Bertha whom I met on the Lineham Trail one summer day in 1994. She was old even then. Her belly sagged and her jowls flapped as she moseyed up the trail, head down, toward me and my hiking companions.

We raised our arms and shouted to get the grizzly's attention. Thirty metres away she finally stopped and looked at us. She showed no surprise or aggression; if anything, she looked exasperated.

After assessing the situation, the old bear stepped off the trail and quietly picked her way through the deadfall, passing only 10 metres away. She vanished around a corner and climbed back onto the trail – right into the midst of a cluster of little girls and their camp counselors. The quiet summer afternoon erupted into a chorus of terrified shrieks. By the time I ran up the trail to help, the bear – no doubt a nervous wreck by now – had vanished. The counselors got the girls under control, and they all hiked down the trail, wide-eyed with excitement.

We all expected the worst from that bear. But all she wanted was to go up a valley too full of humans. We were the problem, not her; she solved it peacefully in spite of our fear.

The grizzly's neutral acceptance of humans, and her age, led me to conclude that this bear might be Bertha. Bertha was one of the first bears biologist David Hamer radio-collared during a 1970s research study. Throughout the study she treated humans as little more than scenery. Her personality couldn't have been better suited to a grizzly bear living in a crowded national park.

Bertha, however, didn't spend all her time in a national park. Like most grizzlies, her home range was larger than the 525 square kilometre Waterton park. In early 1997, a hunter shot her as she wandered through the ranching country north of the park looking for cow afterbirths and new greenery. She was at least 26 years old.

The media took quite an interest in Bertha's death. The debate over whether Alberta should continue to allow grizzly hunting was raging loudly in Calgary and Edmonton. Since Bertha had a name and a history, her passing somehow seemed more poignant.

Biologically, however, Bertha's work was done. The elderly bear could no longer produce offspring. Her teeth were so loose that one virtually fell out in the hand of a wildlife officer examining her carcass. Bertha had been due to die.

Most game biologists estimate that grizzly populations can sustain a conservative loss of maybe 4.6 percent each year to all forms of human-caused mortality, including hunting. If 20 grizzlies range south of Pincher Creek (Alberta Environmental Protection estimates 35 or so), and hunting is the only unnatural source of losses, then the population could withstand Bertha's loss. Nobody, however, knows how many grizzlies live in the area.

The argument about whether hunting is a threat to grizzly numbers will not end until somebody figures out how to count bears. That may happen soon. In 1997, biologists from B.C., Alberta and Parks Canada began to inventory grizzly bears using "hair traps" around scent-bait stations. When grizzlies investigate the foul-smelling baits, they leave tufts of hair on strands of barbed wire strung around the bait. Biologists use the genetic material in each hair sample to identify individual bears. When the study ends, they will have a more accurate estimate of grizzly bear numbers than before.

Meanwhile, provincial biologists try to take a biologically cautious approach to grizzly hunting. They estimate bear populations conservatively, then set quotas, again conservatively, on how many bears humans can kill or remove each year. If wildlife officers haul away any grizzlies to protect livestock and property, they subtract those from the number available to hunters. And, to protect breeding females, they forbid hunters to shoot females accompanied by cubs, or any bear in a group of two or more.

Bertha died in 1997. In 1996, a hunter shot a large male on a nearby ranch. Those were the only legal hunting kills south of Pincher Creek. However, wildlife officers trapped and removed at least six other grizzlies in 1996, and fourteen so far this year. These bears had been attracted to the rotting carcasses of calves that died naturally, or to poorly stored animal feeds. At least four were mature females. Not only did the population lose these bears, it lost all the cubs they might have produced. For an animal that produces so few young, that's a big loss.

Ironically, grizzlies suffer even worse when ranching surrenders to other land uses. In the foothills around Okotoks or Cochrane, rural residential developments and small acreages have replaced the big ranches where grizzlies used to range. Grizzlies can't stay out of trouble when people proliferate across the landscape. That's why those areas no longer have any grizzlies at all. When ranchers sell to developers, grizzlies lose habitat permanently.

Most ranchers in grizzly country never have any problems with grizzlies because they don't invite trouble. They clean up the carcasses of stillborn calves and other animals that die on the range, store animal feed and their garbage safely, and know where bears are likely to be and how to avoid them. "It's always the same guys who have bear troubles," one rancher said to me. "Sort of tells you something, doesn't it?"

Alberta's foothills ranching country still has the potential to sustain good grizzly numbers, partly due to the continued survival of large ranches and the good stewardship of many ranchers. For that reason, educating and motivating the rest of the ranching community to adopt more bear-friendly management practices may do more for grizzlies in the long run than banning the spring grizzly hunt.

Whether to allow grizzly hunting is partly a biological question, partly a moral question and partly a socio-political question.

Biologically, grizzlies are among the slowest-reproducing animals in North America – mostly because other animals don't eat them. They lack the biological characteristics of a prey species – high breeding rate, ability to quickly disperse and recolonize new ranges, and high population densities.

Even so, the deaths of a few male bears may not hurt the population – if there are enough bears and their death rate from other causes is low enough. If biologists know how many grizzlies are in a population, how fast they reproduce and how many die of all causes each year, they can judge if it is safe to sell any grizzly hunting licences. Until they know that, the biologically sensible thing is to err on the side of caution and sell few or – as many now argue – none at all.

Morally, it's hard to justify shooting an animal you do not intend to eat, which has done you no harm and that the population may not be able to spare. But in a pluralistic society based on principles of freedom, moral choices are generally left to the individual. As a hunter, I must constantly test the morality of my choices in the field. Personally, I believe that killing a grizzly would be morally wrong; I never will. Others see things differently. Whether to allow grizzly hunting is a decision that should not be based on one group's moral position. Imposing morality at the expense of freedom of choice is something many who oppose hunting feel strongly about in other contexts.

The socio-political side of the question is perhaps the most challenging. If government bans grizzly hunting, what will this say to rural residents who feel that the current hunting season keeps bears wild and controls their populations? That society as a whole has no interest in managing grizzlies, even though ranchers and other rural residents must continue to live with the bears and their potential dangers? If that is the nonverbal message, how many grizzlies will quietly – and illegally – vanish from isolated pastures as individuals take matters into their own hands? "Shoot, shovel, shut up" say the bumper stickers in Montana, where the U.S. Endangered Species Act rigorously protects grizzly bears. Some biologists speculate that closing the grizzly season in southern Alberta could result in the undocumented and illegal deaths of more bears than would be saved from hunters' bullets.

In the continuing debate over how to conserve wildlife, hunting has always been an easy target. Most people don't hunt. People who don't hunt find it hard to believe that anyone who consciously chooses to kill animals honestly cares about wildlife. If you ban hunting, the simple logic goes, you save animals. That's what we want to do, right?

But dying is part of nature. In any population of wild animals, it isn't how animals die that matters, it's how many. Right now, in southern Alberta, more grizzlies die because of illegal kills, road kills and control removals, than legal hunting. Limiting any of the former will do more good than banning the latter.

In any case, what matters most to a grizzly is whether there is habitat worth living in, not how it will eventually die. The odds of postponing its inevitable death increase in direct proportion to the quality of habitat: roadlessness, wildness, naturalness. As another bumper sticker declares: "It's the habitat, stupid!"

Grizzlies are safest in wilderness; we need more wild habitat. In fact, we need to think seriously about taking some habitat back from industry and restoring it to wildness. But grizzlies can also live – albeit less securely – in areas used for ranching and logging. The key is for enlightened and ecologically literate humans to make the right choices – remove garbage, clean up livestock carcasses, close roads and keep grizzlies from coming into conflict with people. We need to focus a lot more effort on this part of the conservation equation.

Grizzlies have never persisted long in a fragmented landscape full of human development: look at Okotoks, Cochrane or even Banff. Ultimately, the incremental subdivision of the rural west is the biggest threat facing grizzlies and other wildlife. I can't help thinking that energy used on the fight to ban grizzly hunting would be far better spent on the more difficult, and far more important, challenge of keeping human populations and human use as low as possible in bear country.

In any event, a decision to ban grizzly hunting will unquestionably work against the survival of the great bear if it becomes a sort of environmental placebo. It simply is not true that if hunters can't kill grizzlies, then the bears will be safe. The real grizzly conservation priorities are far more complicated than simply locking up hunters' guns. The most important work must involve protecting scenic public lands from real estate promoters, stopping the insidious conversion of ranching country into increasingly cluttered complexes of acreages and rural residences, making ranching more bear-friendly and keeping roads and industry out of our last wild places.

C H A P T E R 5

CHEVIOT
A LAND WHERE COAL IS KING

83C 26 7 80

you would like to say
either you see them or
you don't. in the backed-up
pools above Cadomin

where we took the kids
to fish, trout break
surface with articulate

hunger, the silver
arc that falls
into disappearance

unless it happens where
you have waited
and even then

there is only patience
rippled water, the kids
shouting, look, they're jumping

over here and you know
you are too slow for
anything but absence

the sun steady on
ledges of granite
your line cast

into cold mountain
water, running into the quick
silver

- by Monty Reid[1]

— INTRODUCTION —

Cheviot – no word uttered in the past several years better summarizes the chasm separating the champions of resource extraction from those who equate landscape preservation with alternative ways of earning a living or heritage values. In the history of Alberta's environmental controversies, Cheviot looms large. It is as prominent in the late 1990s as the Oldman River Dam or the Alberta-Pacific pulp mill controversies were a decade earlier. Like those earlier battles, Cheviot is a classic "development versus preservation" debate. Over the course of its twenty year life span, the Cheviot mine will strip nearly 70 million tonnes of coal from the mountains 70 kilometres south of Hinton. It rivals the oils sands expansion proposals in Alberta's northeast for the magnitude of its landscape impact. With a permit area roughly 23 kilometres long and 3.5 kilometres wide, Cheviot's "footprint" is extra-large – nearly three-quarters the size of the City of Vancouver.

For the regional economy, an economy that has relied considerably upon coal mining throughout this century, Cheviot is a lifeline – a project that, as long as the Asian appetite for metallurgical coal stays healthy, would breathe life into coal mining for at least another twenty years. For environmentalists, the miner's lifeline is a noose threatening the world heritage status of Canada's mountain parks, the health of vulnerable species such as grizzly bears, and the prospects for developing alternative foundations for the regional economy.

I.U.

— THREE CHEERS FOR CHEVIOT —

WHOLE TOWN NEEDS TO SUPPORT CHEVIOT[2]

- by The Hinton Parklander

With the public hearing on the proposal for a new Cheviot mine site in the West Yellowhead region fast approaching, now is certainly an appropriate time for us to encourage members of this community – and surrounding areas – to participate in any way they possibly can.

During last fall's meetings between representatives of the Cheviot coal mine project and environmental groups that are trying to prevent the establishment of such a facility it was super to see so many mine workers showing their support at the meeting.

Likewise, it was great to see that their words were able to get shared via Alberta's and Canada's larger media outlets.

With that in mind, we would like to once again encourage participation in the upcoming public hearings – which begin Jan. 13 – from all sectors of Hinton's community, but specifically from the grassroots level.

After all it is the workers at Cheviot – from management down – who can effectively let the public know just how important the mine is to them personally, their families and their friends.

To see an employee from the mine take the time to show his or her support for their company is nice.

There are plenty of workers here.

Let's hope they all get out and help themselves. And Hinton too.

HINTON AND DISTRICT CHAMBER OF COMMERCE: CHEVIOT TESTIMONY

One of the striking features of Cheviot is the widespread support the project received from business, labour, and municipal governments in the Hinton area. The Hinton and District Chamber of Commerce, representing 240 Hinton-area businesses, strongly endorsed the Cheviot project during the federal-provincial review hearings in January 1997. In his presentation to the joint review panel, Bruce Deal argued that the mine should be regarded as a temporary use of the land and the impact of the mine on the environment would be mitigated by the reclamation activities of Cardinal River Coals.

For the Chamber, the Cheviot issue was only a legitimate concern of Hinton-area residents. He said that Cardinal River Coals (CRC) has "mitigated the concerns of interested groups with legitimate concerns; and, again, you know, there's been some discussion, here, about groups that have come to make presentations that are not part of our community; and I guess I question how legitimate their concerns are."[3] No doubt this hostility to outsiders arose from the Chamber's estimates of what the refusal of the Cheviot project would do to the economy in the Hinton region, an argument the Review Panel proved to be very receptive to. The following excerpts from the Chamber's presentation to the review panel summarize the Chamber's gloomy assessment of the negative consequences which turning down the Cheviot project would have upon the CRC workforce and the Town of Hinton:

Regarding the employees of Cardinal River Coals Ltd.:

"• 450 Direct jobs will be lost.
- Many of these people will have to move and start over in another community. They will lose thousands of dollars as house prices fall. Their children will have to start school in new communities, not a pleasant experience for some children.
- Some will live on Unemployment Insurance for some period, their standard of living will drop, their savings will be depleted.
- Some will accept retraining. The Taxpayers will have to bear the cost of that retraining.

- Some will not find another job. The stress of unemployment brings both financial and social problems; spousal abuse, family violence, and substance abuse are sometimes the result.
- The employees of Cardinal River Coals Ltd. have worked hard for those 450 jobs. They deserve those jobs. They deserve another job at the Cheviot mine."

Regarding the Town of Hinton and its surrounding area:

"The Town of Hinton and surrounding area has a population of approximately 10,000 with the Cheviot mine estimated to produce 450 direct jobs, representing 15% of the work force within the area.

These jobs, and jobs generated through multiplier effects produce $31 million dollars of household Income and on a value added basis put $55 million dollars into the local economy. The Luscar mine with its stable markets and workforce provides a major impact on the local economy. Additional Luscar purchases of goods and services contributes another $100 million into the provincial G.D.P.

Without project approval, the loss of such significant dollars into our community will have an immediate and devastating impact on our local economy which will:

- Stop the economic growth which we have been enjoying in recent years.
- Seriously affect the business community resulting in additional loss of jobs in the retail and services sectors and certainly result in SEVERE business downturn with bankruptcies occurring."

Bleak forecasts such as these led the Chamber in its concluding statement to "strongly urge this panel to approve this exceptional opportunity for the Mine workers, the community of Hinton and surrounding area, and the Province of Alberta. Again, the impacts of the Project are not a matter of new jobs and economic benefit; it is a matter of lost jobs and lost economic benefit of catastrophic proportions."

I.U.

"If what I saw at CRC is any indication of what the Cheviot mine practices, CRC is certainly sustainable in my terms. I'm certainly satisfied and convinced."

- Audrey Cormack, President of the Alberta Federation of Labor

THE TOWN OF HINTON: BRIEF TO THE CHEVIOT HEARINGS

Like the Hinton and District Chamber of Commerce and the United Mineworkers, the Town of Hinton strongly endorsed the Cheviot project. Excerpts from the Town Council's presentation, delivered by Mayor Ross Risvold, follow:[4]

I.U.

INTRODUCTION

This presentation is made on behalf of Hinton's seven member elected Town Council. We are the democratically elected representatives of our community. We represent all of Hinton's citizens (last population census in 1994 was 9,341), 87% of Cardinal River Coals Limited's (hereinafter referred to as C.R.C.) workers and their families live in Hinton. They and their neighbours all benefit economically from coal mining and have much at stake in the Cheviot Mine decision. Indeed, the economic social fabric of our entire com-munity gains much strength from the existence of all coal mining in our area...

Our core message is: The Cheviot project represents a sound balance between social development, environmental responsibility, and economic benefits; and must be approved. . .

OVERVIEW OF COMMUNITY

...Based on the responsible development of forest and coal resources in this region, Hinton has experienced very solid growth...

From the Town of Hinton "Social and Economic Impacts of the Proposed Cheviot Mine" study we commissioned Nichols Applied Management to undertake... we quote:

> A number of economic indicators confirm the relative strength of the town's economy. Household incomes, population growth, and labour force participation rates are all higher than the provincial average while unemployment rates are somewhat lower.
>
> Accompanying the economic growth of Hinton over the past two decades has been a steady social maturation of the community. The average age in the community has risen over time, mobility rates have declined, and family stability has increased.

The people of Hinton and area are very proud of where they call home... Our major employers, being the Weldwood pulpmill, HiAtha sawmill, and three coal mines, are considered excellent corporate citizens.

In this context, Cardinal River Coals Limited has been very active in their communication leading up to development of the application for the Cheviot Mine. We believe C.R.C. has demonstrated an openness and constructiveness in the process of involving stakeholders. As a Town Council, we've had ample opportunity to input and comment, and to ensure this coal mine development addresses positively all the issues. I personally get rather frustrated with comments (usually from outside the community) that there hasn't been ample opportunity to understand the issues surrounding the Cheviot project and constructively participate. I also get frustrated when people oppose the project completely and are not interested in the mitigation of environmental concerns.

We want to re-emphasize that the Cheviot Mine is a replacement mine for the Luscar Mine that has been operating successfully and responsibly since the late-60's. Open pit coal mining being done responsibly is not a new thing in this area and the latest management practices are being applied to the plan for Cheviot. Cardinal River Coals has a sound positive track record of environmental stewardship.

With this Cheviot Mine proceeding, socio-economic life of Hinton continues in a positive fashion with no substantive change. Should the Mine not proceed, there will be a devastating effect on our community. This would be from the loss of population, incomes, and confidence within our community.

DETAILED ASSESSMENT (SOCIAL, ENVIRONMENTAL, & ECONOMIC)

a) Social Development

...Coal mining has brought a lot of stability to our community socially and demographically. 94% of CRC employees report they have lived in the area more than 5 years. This compares very favorably to our 1991 census results for the community overall, where 68% of our residents reported to being here more than 5 years. (We are pleased this is up substantially from the 1981 census when it was 57%)...

(Nichols Applied Management concluded):

> It is expected that the Cheviot mine will employ a total of 430 workers during its operational phase. Those workers will transfer to the new mine from the nearby Luscar mine. If the mine does not proceed, those jobs will be lost. With close to 90% of the mine workers living in Hinton, the town will face a reduction of about 390 mining jobs and up to another 240 jobs in various service industries that rely indirectly on the mining activities. Recognizing that many displaced workers will have to relocate to obtain employment elsewhere, it is estimated that the town's population may decline by as many as 1,300 people (almost 400 households)...

> For those workers and their families choosing to relocate under the "no Cheviot" scenario, the community would experience negative social impacts related to 1) a disruption to the affected local families and to the town's overall social fabric, 2) a loss of valued and involved residents, and 3) increased family and marital stress.

The population and social impacts of the Cheviot Project not proceeding are substantial and real!

With a decision to proceed with the Cheviot Mine, our population is expected to reach about 11,250 by the year 2006. Without this replacement mine, over the 5-year period beginning in about 1998, we expect to lose over 15% of our population. Hinton residents do not wish to see any of their neighbours, whether C.R.C. workers or effected retail/service workers, lose their jobs and face extraordinary economic and social pressures. While our collective community heart is big, this would be a tragedy and put incredible social pressure on all.

b) Environmental Responsibility

Hinton residents and Hinton Town Council are concerned about the overall environment. We've reviewed the air (biophysical), soil/terrain/ hydrology, vegetation, wildlife/fishery, and water quality/health risk impact assessments filed in the Cheviot Mine application and have confidence the long-term environmental impacts of this project are generally addressed, and will be minimized by C.R.C....

As we reflect on the environmental record of CRC to date, the actual accomplishments demonstrate a high level of corporate environmental stewardship...

c) Economic Effect

The current Luscar mine, and it's replacement, Cheviot, are a critical part of the present prosperity and future economic health of Hinton and area. (sic) The Nichols Study... provides much evidence in this regard. The economics are big and include the following:

- Cheviot project represents 400 direct jobs – a slight reduction from current levels at Luscar Mine. The is 13% of the work force in the Hinton area and contributes over $30,000,000 to household incomes annually. Add to this the direct local spending by C.R.C. of $8,000,000 annually.
- Over the 1997-2004 period, there will be additional construction and reclamation/mine decommissioning activity benefitting directly local contractors, service industry and others.
- Overall, the replacement of Luscar mine with Cheviot will economically sustain current levels of overall community income...

While Town Council did so with reluctance, we analyzed the negative implications on Hinton economically if the Cheviot mine does not proceed. The results are serious. There would undoubtedly be a real financial, and strong emotional, brake put on most economic elements of our community for some time. The effects include:

- A population decline of almost 400 households.
- Residential real estate prices would likely go down 20-30%. Commercial/retail owners would lose about 15% on current property values.
- There would be a loss of direct local spending by C.R.C. of $8,000,000 annually. When this is added to the portion of C.R.C. employees payroll spent locally, the cumulative annual household income in Hinton drops 15%. Perhaps another 240 jobs would be lost in various retail and service businesses that rely indirectly on the mine and mine workers.
- A municipal property tax shift of approximately 12.5% or $700,000 onto in-Town Hinton industry, being the Weldwood pulpmill and HiAhta sawmill, due to the reduced residential and commercial property values in Hinton (sic).
- Volunteer organizations reliant on user revenues such as the Curling club, Golf course, Legion, etc. would be required to either downscale services or raise fees to remaining customers.

We have not quantified the economic cost of the resulting social disruptions to the community, including retraining, marital/family crisis prevention/counselling and support, savings depletion, etc. Indeed, even now as we await the approval of Cheviot mine, the survey of CRC workers revealed over half the workers avoiding some major financial commitments or delaying major purchases.

I don't like to dwell on the negative very long. The Nichols Study summarizes the economic effect well:

> The development of the Cheviot mine will generate a positive but relatively modest impact on the town's economy during the period 1997 to 1999. The major impact of the mine will derive from its long-term operation, which essentially will maintain mining activity in the area near current levels.

CONCLUSION

... The Town Council of Hinton and the vast majority of Hintonites support Cardinal River Coal's Cheviot Mine Project (sic). A decision to proceed expresses well placed confidence on the Cheviot project's overall soundness. It also reaffirms community confidence in our collective, responsible future. The project is sound on all bases, social, environmental and economic. It is a balanced project. We urge you to approve the Cheviot Mine Project.

"I will never apologize for defending anyone's job."

- Audrey Cormack, President of the Alberta Federation of Labor

CHEVIOT MUST BE OKAYED[5]

- by The Hinton Parklander

It is certainly within the best interests of West Yellowhead residents to consider the future of the Cheviot coal mine – and industrial development as a whole – when decisions are made about the political future of this province.

Members of the New Democratic Party, represented by Glenn Taylor in this riding, have stated that they support the successful establishment of the Cheviot coal mine.

West Yellowhead MLA Duco Van Binsbergen has spoken out about his support for the mine as well.

Certainly the progressive attributes that go along with another mine being built in this healthy province is something the extremely popular Conservative Party would also look kindly upon.

So, in a nutshell, it's difficult to find any decision makers who would go against the construction of Cheviot in this region.

For those reasons we must once again say that members of the review panel – those who are looking into the concerns relating to the mine – should consider the future of residents living in the region very closely when decisions are made.

Environmental groups have raised interesting concerns. Water, wildlife and trees have been the topic of several discussions.

We believe Cheviot cares about the environment.

And we don't think our children will suffer because of the company.

— VOICES IN THE WILDERNESS —

DOWN THE ROAD TO CHEVIOT

- by Ian Urquhart

I could not have ordered a more glorious morning to drive south on Highway 40 from Hinton to look over the site of the Cheviot mine – the mountain landscape Cardinal River Coals (CRC) covets for its 70 million tonnes of coal. Although it was early, the sun had taken advantage of the cloudless sky to propel the thermostat into the mid-teens. It wasn't long before I saw deer – first whitetail, then mulies. The air on that late May day tasted fabulous – certainly better than what I had sampled earlier in the vicinity of the Weldwood pulp mill at Hinton. It was filled with the calls of songbirds, the drumbeats of grouse, and the aroma of pine – all coming from the second-growth lodgepole forests planted by Weldwood over twenty years earlier. Standing in that forest I could be convinced that the environmental consequences of heavy industrial use in the Rockies and the foothills could be quite benign.

It wasn't long before my happy thought was put to the test. Once you are in the vicinity of the area's active coal mines – Gregg River and Luscar – the heavy duty machinery you spy soon outnumber the deer and sheep you are likely to see. The offspring of Hitachi, Volvo, and North America's very own Caterpillar, could be counted among the mammoths moving earth – either to build roads or to uncover coal.

Set against the backdrop of the Rockies the mines are rather Lilliputian. Their processing facilities, although many stories tall, are puny against the slate grey front ranges of the Rocky Mountains just a few kilometres to the west. The imagery of the mines is cold and stark. If you did not realize it before hand, strip mining operations are not subtle. At Gregg River, empty rail cars inched towards the loading area, screaming from the metal to metal contact of wheels to track. From the mine's twin towers, decorated with two rows of huge discarded haul truck tires, the whirring and chugging of machinery supply the harmony for the rail cars. They force your ears to strain to hear the warblers in the forest across the road from the plant.

Further down the road at the Luscar viewpoint you appreciate that there is nothing Lilliputian about the impact of strip mining upon the landscape. Across the valley from the viewpoint, mountain sides have been cut open with surgical precision so that scurrying haul trucks can deliver ton after ton of coal to the always hungry processing plant. In this setting big technology rules.

Yet, if you look at the reclaimed hillside behind your back you may see one of the ironies of strip mining. Not 150 yards from the viewpoint and well within sight and earshot of the mine I counted nineteen bighorn rams, many with massive full curl horns, lounging on the grassy hill. It is this latter scene which CRC and its supporters use to lend credence to the claim that, with time and effort, some species may adjust and adapt well to open pit mining.

For me, the scene was more reminiscent of Monty Python's "Flying Sheep" sketch. Like Harold, "a clever sheep," the rams behind me realized that "a sheep's life consists of standing around for a few months and then being eaten" (by wolves, or shot by hunters). Luscar's sheep know that, as long as they stay close to the mine, they are safe from both types of predators. What must impress many passersby as a sign of a healthy, natural environment is in reality very unnatural, an oddity.

Once through the hamlet of Cadomin, the Grave Flats Road (for environmentalists, an ironic name) takes you to the proposed site of the Cheviot mine. The word "site" doesn't do justice to the Cheviot proposal. It doesn't capture its enormity, its gargantuan scale. During the course of Cheviot's projected twenty year life span, CRC intends to use open pit mining techniques to strip coal from an area fourteen miles long and roughly two miles wide.

The Mountain Park townsite sits near the heart of Cheviot and is not much more than a stone's throw from where the coal preparation plant will be built, near the base of Harris Mountain. Up until the 1950s Mountain Park had been a vibrant coal mining community. But, when the mine shut down, the town died, the bulldozers came, and all of the buildings were torn down. The only indications that Mountain Park once was home to as many as 1,500 people are the cemetery and a few handmade interpretive signs.

It was easy to see why that earlier generation of coal miners had chosen Mountain Park for their home. Its alpine setting is spectacular. Mountains surround the townsite to the west, south, and east – Cheviot and Tripoli Mountains erupt out of the Miette Range to the west/southwest while Harris Mountain dominates the view to the east. To the north, the eye is drawn to the headwaters of the McLeod River as they rush down the valley towards Cadomin.

The day I discovered Mountain Park I did not find much solitude. As the townsite came into view, so too did four travel trailers, a camper, a handful of quads, and five young families. Parents and kids explored the townsite. As two mothers, escorted by four preschoolers, approached me at the site of the old train station the streetproofed escorts warned that "a stranger in a black truck was up ahead." I asked them what brought them to the townsite. For the moms and their children this visit was their first. Although Hinton was home, they had never been to Mountain Park before. Usually the men made the trip alone – bringing horses instead of quads – and spent their time riding further into the backcountry.

This year was different. The federal and provincial governments had approved the Cheviot project and the only thing stopping CRC from starting construction were last ditch court cases from a native band and a coalition of environmental groups. "This year," one of the women offered, "our husbands thought all of the families should come. It will probably be the last chance we have to see Mountain Park like this."

On the one hand, after all I had read about how important it was to the people of Hinton for Cheviot to proceed, I was glad to meet some locals who valued Mountain Park for something other than the coal which lay buried beneath our feet. But, as a friend pointed out to me later, wasn't it more ominous that these people (and he might as well have included me too) had waited until Cheviot to discover this landscape. Their ignorance, my ignorance, is troubling for what it says about our hierarchy of values, the difficulty I believe many Albertans have in exploring seriously uses other than mining, logging, or petroleum production for resource-rich public lands. Our eleventh hour appearances suggested how much longer it takes and more difficult it is to regard landscapes like Mountain Park as places which should not be earmarked for whatever economic endeavour will produce the greatest profit, the greatest number of jobs, in the shortest period of time.

As we have seen from the preceding excerpts from briefs and testimony heard by the federal-provincial review panel which considered the Cheviot project, the coal which lay beneath our feet that day is crucial to the economic and social stability of the Hinton region. The workers who will operate the trucks, shovels, and processing facilities at Cheviot will be the workers who perform those jobs today at the Luscar mine.

Having grown up in Trail, a community which owes its existence to mining, these arguments are powerful ones for me. What would have happened to my dad's business, to my family, if the smelter had run out of lead and zinc to refine? Had the smelter died we would have scattered – maybe across the province, maybe across the country.

Yet, as you will see below in the compelling language of Mike Bracko, places like Mountain Park cannot be said to belong to only one group in society. The claims of other Albertans and other Canadians to share in the decision making about the future of the mountain setting CRC covets are just as legitimate and compelling as the views of workers whose future employment in the region depends upon finding more coal to mine.

For several days after I climbed into my truck and came home to Edmonton I tried to come to terms with what I had seen on my trip to Cheviot. Why, as my friend challenged me, would I idealize the coal miners of Mountain Park and not the coal miners of Luscar?

The empathy I feel for an earlier generation of coal miners and for the people today who want to preserve thc townsite is not just nostalgia. Part of it comes from the dramatic differences in the landscape impacts of coal mining then and now. Today, with the power of Big T at their fingertips, coal miners smash mountains to get at their treasures. The underground coal mines of yesterday, whatever their shortcomings, simply didn't chew up the landscape like strip mining does. Technology has changed the environmental consequences of work. When it comes to the surface landscape impacts of coal mining, technological change has not been benign.

More importantly, for the miners of sixty or seventy years ago Mountain Park was more than a place to work – it was home too. And, as home, Mountain Park and its surrounding peaks and valleys assumed an importance and a meaning that Cheviot will never have for the workers from Hinton who will earn their paycheques there. In the Mountain Park cemetery you find a brief history of the town, a history suggesting that the people who lived there cared deeply about the surrounding landscape. It's that ethic which is missing from the bosses and miners of today, despite their commitment to reclaim the landscape. Linda Goyette's column outlines that ethic. So too does Mike Bracko, a coal miner who grew up in Mountain Park, when he says that he doesn't want CRC to "fill the valleys... and destroy everything I believe in." If I too idealize Mountain Park, it's because of the richer, multidimensional relationship the people who lived and worked there forged with that landscape. It's the type of relationship we should strive for.

MIKE BRACKO: THE ESTEEMED ELDER

Mike, "Spike", Bracko represented the Mountain Park Environmental Protection and Heritage Association at the federal-provincial hearings concerning the Cheviot Coal Project. At 73 years of age, Mr. Bracko was described by one of the opponents of the coal project as an "esteemed elder. . . whose leadership everyone here would be wise to follow." Born and raised in Mountain Park, the townsite located at the heart of the Cheviot Project area, Mike Bracko spent his entire working life, 47 years, in the coal business and retired from Luscar in 1987. During the public hearings, his testimony stood out for here was a Hinton-area resident who broke ranks with his community and questioned the Cheviot project. Although he said that he was not against Cheviot per se, he opposed the company's plans "to fill the valleys... and destroy everything I believe in."[6] At the hearings his Association asked that the entire former townsite be left undisturbed and that there should be less surface disturbance in the area around the former townsite.

The following remarks are taken from his commentary on the presentation which Mayor Ross Risvold made on behalf of the Town of Hinton to the public hearings in February 1997.

I.U.

"I'm representing an association and they have backed me up 100 percent. This is a group of people from Mountain Park that want to try and preserve our Heritage and everything else up there and we can't get anywhere, Ross.

You know, you can say all these good things about Cardinal River, we had a million meetings with them, whatever, but we, as people of Mountain Park, can't get anything from them. We can't get anything. And that's our problem.

The environmentalists that come here and sit here, they're from Calgary, they're from Edmonton, and I read in the paper where why are they come in here not minding their own business, minding the business of people in Hinton?

Ever since I come back here from the prairies, I have always thought – I went to Mountain Park and I felt Mountain Park was mine, mine. This is where I grew up and this is where I was raised and this is where I walked the trails, I go to the cemetery, this is where I got relatives, all this stuff. I felt Mountain Park was mine.

Then as time went on, the times I would go up there, these dirt bikers would come up there, their quads and whatever, rip and roar and tear up the places that I thought was unbelievable, beautiful, that the good Lord had put there for us.

And anyway, I was mad. It used to really make me mad and then I would quit going. The only days I would go would be middle of the week or something so there's none of them up there. They're all home in Edmonton or wherever they're at.

But then you get to realize, God darn it, I'm just one person. Canada is a big country, there's a lot of Canadians here. They're entitled to go up there. Albertans, they're entitled to go up there and so on and so forth. Ross, I accepted it. I had a hard time, but I accepted that, and now we got publicity coming out of our ears.

Our local paper, there isn't one you can pick up that there isn't somebody within the community crabbing about these environmentalists are coming here. These people are, for all we know, are representing Albertans someplace else, Albertans, not Hintonites.

I didn't want the bikers – to share Mountain Park with the bikers. Here everybody in Hinton is trying to claim Mountain Park now. No, a few years ago they scoffed at Mountain Park, but now they want to claim it. And I think that's wrong.

Anybody in Alberta, everybody in Alberta owns a piece of that Mountain Park and a piece of that – and a tonne of that coal that's under the ground or whatever and they own a piece of the valleys, they own a piece of the creeks, they own a piece of everything and I personally don't see any company, Cardinal River or anybody else, going up there and destroying all that stuff. I don't.

Mountain Park is a precious place to me and I have come here and sat every day trying to get the opportunity to say something, express my feelings. I know I'm not too good at it, but God darn it that's the way I believe and I'm going to say that."[7]

LIFELONG BONDS TO A VALLEY[8]

- by Linda Goyette, The Edmonton Journal

Kay Farnham and her friends had no chance to speak at the Cheviot mine hearings this winter. That's Alberta's loss.

They have an intimate knowledge of the mountain valley on the eastern edge of Jasper National Park. I think they qualify as expert witnesses as much as mining executives and environmental scientists.

The Alberta Energy and Utilities Board and the Canadian Environmental Assessment Agency are expected to rule on the future of the $250-million project within two weeks. They will reach their decision without the benefit of the following testimony. Let's put it on the record, anyway.

"Every little bit of that valley that goes is a loss to the world," Farnham begins. "The world will finally run out of resources, and what will be left of it?"

She is an environmentalist – "absolutely, I certainly am" – but she doesn't quite fit the public image of the mountaineering ecologist.

Farnham is 90. Back in 1919, when she was an adventurous 13-year-old, she moved with her family to a wilderness mining camp called Mountain Park. She fell in love with the wild beauty of the place. Growing up in the "highest town in Canada," she became a local schoolteacher.

"We had a tremendous amount of freedom, and fun and good times," she says. "Oh, to feel that free in a lifetime."

The Albertans who once lived in the ghost towns of the Coal Branch – Mountain Park, Cadomin, Luscar, Mercoal, Coal Valley and Leyland – are exiles from a lost way of life. The underground mines closed in the 1950s, the isolated communities were bulldozed to nothingness, and the mining families scattered across North America.

Yet they have a way of finding one another. They organize reunions, and return to the valley often on their own because they can't stay away.

Once a year Farnham has lunch with her former Mountain Park students. Most of these women are now in their 70s. On Friday, a group of 21 met again to tell stories, and share their photographs, poems and pages of handwritten reminiscences about life in a spirited community.

Well-informed about the Cheviot mine project, they also talked about their concerns for the valley.

"You wonder why there's such a bond between us?" says Mary Liviero. "Well, when one of us talks about the forget-me-nots on Miner's Hill, the rest of us know exactly the place. The environment is a very important part of our bond."

They lived in complete isolation at the end of the railway line – the first dirt road to the outside world wasn't built until 1934 – and so the natural world became their larger home.

"We hiked up and down those mountains all the time," said Caroline Russell in the handwritten story she gave to me. "That centre hill was covered with fireweed. On the way to Mount Cheviot were the columbines, snow flowers, lady-slippers..."

"Even though life was a little difficult at times, we never felt alone or neglected," added Cecelia Dmytryshyn in her own memoir. The women can describe every creek, every cliff, every kind of wildflower, as if they'd explored the valley last week.

Cardinal River Coals wants permission to build a series of 26 open pits in a strip 3.5 kilometres wide and 22 kilometres long, just outside the national park boundary. The project would create 900 construction jobs and preserve 450 mining jobs.

Conservationists oppose the project because the area is a migration corridor for wildlife. They say the area also deserves protection because it contains rare landforms, plants and insects that have evolved without interference for an estimated 20,000 to 40,000 years.

"When I first heard about the mine project I cried," said Regina Dotto. "They will be destroying our territory. I need to go back there, and see it the way it was."

"An open pit mine will destroy a part of our bond, a part of our heritage," adds Liviero. She wonders how long the mine will stay open when coal-buying countries can find cheaper sources of coal in poorer nations. She objects to the company's determination to mine the coal in open pits, rather than underground mines, without giving Albertans enough time to consider environmental consequences.

"It's the greed that gets to me. It's the plain, old greed."

LaVerne Dunlop and Olivia Richards said they've investigated reclamation work at Luscar, and they don't believe the company restores the landscape properly. "They're destroying the area," said Dunlop. "They take the feeling away when they take the mountain away."

Nora McKay believes the Cheviot mine will be approved and built. "But I hope they spend more money, and use more discretion, in the way they mine the coal. That project is not the way."

Ready for the worst, Gloria Dagil has only this to say to Luscar Ltd.: "They'd better not mine Third Hill, that's for sure."

As a young child, Dagil scrambled to the top, planted a hand-made cross in the dirt and rocks, and carefully placed some salami inside a bottle as a permanent marker. "I claimed that hill for Queen Isabella of Spain," she recalls dryly.

Their fathers, brothers and husbands were miners. They understand the support in west central Alberta for the Cheviot project – and sympathize – yet they ask Albertans to also consider the gift of nature.

"I do know men need work," said Kay Farnham, "but isn't there another way?"

"It reminds me of a park."

- Orville Cook, Secretary-Treasurer of the United Mine Workers of America Local 1656, on Cardinal River reclamation efforts.

STRIP MINING PARADISE[9]

- Dianne Pachal

Grandpa won't you take me up
to the Mountain Park wildlands,
Up by the Divide where paradise lay?
Well I'm sorry my grandchild,
but you're too late in ask'n,
Mr. Lougheed's coal train
has done hauled it away.

-Re-worded from John Prine's song "Paradise" lamenting the demise of Muhlenberge County, Virginia. Hinton area residents and conservation groups alike worked during public hearings in January and February 1997 to stop this from becoming a lament for the Cardinal Divide wildlands.

The Cheviot coal mine would be perched astride the continental divide between the waterways of the Arctic and Atlantic Oceans, a short 2.8 kilometres from Jasper National Park, high in the Rocky Mountain front ranges southwest of Hinton. The mine proposal was rammed through the federal-provincial regulatory approval process. The mine's proponent, Cardinal River Coals (CRC), expected to receive its mining permits in May 1997, before Canadians had a chance to work out a solution, as was done for the Canadian gold mine that threatened Yellowstone National Park.

CRC plans to dig an open-pit coal mine consisting of 26 huge pits and valley-filling piles of waste rock stretching 23 kilometres long – the width of Edmonton – and up to three and a half kilometres wide. Roughly 25 tons of the mountain landscape will be excavated for every ton of coal removed and much of it dumped onto the surrounding landscape. Destroyed will be eight creek valleys which form the headwaters of the McLeod and Cardinal Rivers.

Cardinal River Coals is a joint venture company consisting equally of Luscar Ltd., chaired by Alberta's former Premier, Peter Lougheed, and Consolidated Coal, the United States' largest coal mining company. CRC's application was reviewed at a joint federal-provincial public hearing held in Hinton. Two of the three-member hearing panel were members of the Alberta Energy and Utilities Board, where history has shown that turning down a development on environmental grounds is extremely rare.

World Heritage Site Threatened by Mine

The United Nations has declared Jasper National Park, together with Banff, Yoho, and Kootenay National Parks, to be a World Heritage Site. This designation means its wilderness and wildlife are of outstanding global significance. At the hearings Parks Canada, the steward of our national parks, concluded: "Clearly, the Cheviot Mine Project, as proposed, has the potential to adversely impact the ecological integrity of Jasper National Park. Parks Canada's concerns relate specifically to the loss or alienation of habitat, impacts on essential wildlife travel corridors which link Jasper National Park and the high-quality habitat in adjacent provincial lands, increases in wildlife mortalities and the cumulative impacts of this project and other planned or proposed activities such as timber harvesting, access, and oil and gas exploration on Jasper National Park."

Dr. Stephen Herrero, a carnivore expert, concluded that the Cheviot mine, when added to the effects of industrial activity already occurring or approved elsewhere in the region, could, over time, result in the extirpation of grizzly bears, wolves, wolverines and cougars from the region. Dr. Herrero was contracted by the company for their mine application. The Alberta Fish and Wildlife Division recently said that if the U.S. were to request some grizzly bears to assist with its recovery efforts, Alberta would have none to give. The Alberta numbers, at roughly 800 bears, are still far below the target of 1,000 set in 1984 as the goal for recovery.

The Fish and Wildlife plan for grizzlies stipulates: "The survival of the grizzly bears in Alberta primarily depends on preservation and management of habitat. A priority for management here [in the Rocky Mountains where about half of the remaining grizzlies live] is habitat inventory, protection and enhancement." The Cheviot mine area is identified as part of the habitat needed for the recovery of Alberta's grizzlies, but Alberta has no endangered species legislation to protect them.

Damage from Mine to Extend Beyond 100 Years

Both the company's and Parks Canada's experts concluded that the mine would result in the direct loss of quality habitat and wildlife travel routes for at least 100 years. At the beginning of their presentation Parks Canada noted "Species such as the grizzly bear are quantifiable measures of ecological integrity and are used as surrogate or umbrella species by which the impact of other species might be assessed. Parks Canada supports this approach, but we must remember that it works both ways." Given that the mine would have significant, adverse effects on grizzlies for at least a century, "[this] suggests a parallel situation for other species under the umbrella."

A Hot Spot of Biological Diversity

There is a preponderance of rare, disjunct and threatened species in the proposed mine area, including some species that have not been found elsewhere in the world. Based on scientific studies of the area's insects and plants, it is believed to have been a biological refugium during the last ice age – a place that became a refuge for life because it was spared the ice cover of the advancing glaciers. A rare butterfly, *Boloria improba*, was first discovered here and a previously unrecorded species of winglessfly has also been found. This is the only location south of the Arctic Circle where some species of mosses are found. Scientist agree there is a high potential to discover more previously unrecorded or rare species throughout the mine area. Twenty-seven species of mammals and birds listed as being in trouble in Alberta also reside in the proposed mine area. They include wolverines, grizzly bears, norther long-eared bats, and Harlequin ducks. Although Alberta still does not officially list threatened plants or fish, the bull trout is a threatened species, and there are plants from fifteen provincially significant and three nationally significant populations that would all be lost.

The Alberta government zoned the proposed mine area as Critical Wildlife – "crucial to the maintenance of specific fish and wildlife populations." It is an island within a narrow strip of Prime Protection zoned lands along the boundary of Jasper National Park. The rest of the region is zoned for development and consists of roads, mines, clearcut logging, and oil and gas activity.

A small provincial Natural Area protects only a portion of this unique place. Scientific assessments recommended the entire Cardinal Divide area be designated as a Natural Area of Canadian Significance. Not only has Western science pointed to the irreplaceable value of the area, so too has the traditional knowledge of native peoples. There is an ancient, unbroken chain of medicinal use of plants from the proposed mine area – plants that cannot be found elsewhere.

Silencing the Spring

Canadian Wildlife Service experts testified that the proposed mine area has a species richness and diversity of song birds "as high as it gets in North America." Many species are declining in North America, including those that nest here in Canada and winter in the southern U.S., Mexico or the tropics. The experts estimated that the mine would result in the loss of 4,000 to 5,000 song birds and their offspring from then on, including birds from thirty-two species whose populations are already declining.

Harlequins – The Spotted Owls of Mountain Streams

The dramatically-coloured harlequin ducks are long lived and return year after year to their same breeding and wintering areas, making them good indicators of environmental quality. They depend on fast flowing, undisturbed mountain streams for nesting and rearing their young. Their numbers are also declining in western North America. The streams of the proposed mine area are home to the second largest known breeding population of harlequins in western North America. Researchers concluded that the only option for the long-term viability of the harlequins was to leave all of the streams of the proposed mine area untouched. Mining plans will turn these streams into open pits or dumps for excavated rock.

There Are Alternatives

The controversy over the Cheviot mine project is not an issue of jobs versus the environment, but one of short-term jobs from a non-renewable resource versus long-term jobs and sustainable development.

This is not the last deposit of metallurgical coal in the region, let alone in Alberta. Other coal deposits are held by other companies also planning new mines or expansions in the Hinton region. There are alternatives which would not threaten the World Heritage site and would not destroy the Cardinal Divide area, including other alternatives for CRC.

The company had planned a mine for its East Cadomin lease near its existing Luscar mine. Substantial coal reserves also remain at their Luscar mine, accessible by both conventional underground mining and their inactive hydraulic mine. Alberta's Coal Conservation Act was supposed to protect the public from companies digging up public lands only to take out the coal cheapest to reach and then abandoning the rest to move on to a whole new area to dig up.

The company acknowledged that it can only predict the coal markets for three years beyond the opening of the Cheviot mine. It's simply ludicrous to destroy the Cardinal Divide area, its wildlife, valleys and streams, for what in the end may only be three years' worth of coal shipped to Asia. CRC admitted that their work force would not be without hope of other employment if the Cheviot mine was not approved. Rather, there will be a strong job market for the workers in 1999 when the mine is planned to open due to other mine expansions and the huge oil sands developments which also use open-pit mining.

The government's landuse planning identified the need to diversify the region's resource extraction based economy and highlighted tourism as an option. The wildlands of the proposed mine area were identified as one of the two key tourism assets in the entire region. The Cheviot mine will foreclose that option.

THREATENING A WORLD HERITAGE SITE

In 1972, the United Nations adopted The Convention Concerning the Protection of the World Cultural and Natural Heritage. The objective of the convention was to define and conserve cultural and natural sites "whose outstanding values should be preserved for all humanity".

Twelve years later, Canada's Rocky Mountain Parks were designated a World Heritage site, a distinction that, as of January 1998, had only been bestowed upon a total of 114 natural sites in the world. Our mountain parks were added to the World Heritage list "in recognition of the area's outstanding natural beauty, floral and faunal diversity, and for being a prime example of ongoing geological processes such as glaciation and canyon formation."

In the eyes of the United Nations, the Cheviot project, given its close proximity to Jasper National Park, threatens the status of the mountain parks. The World Heritage Committee, in a letter to the Canadian delegate to the United Nations Educational, Scientific, and Cultural Organization, asked the federal and provincial governments to reconsider their approval of the Cheviot Project. The letter followed the December 1997 deliberations of the World Heritage Committee which said the following about the emerging situation in the Alberta Rockies: I.U.

"Canadian Rocky Mountain Parks (Canada)

The Committee noted with concern the potential threats to the integrity of this site due to the proposed Cheviot Mine Project, designed to exploit a large, open-pit coal mine, located 1.8 km from the Jasper National Park portion of this World Heritage area. Despite the fact that during the environmental assessment process conservation organizations and Parks Canada expressed concern regarding the negative impacts, e.g. loss or alienation of wildlife habitat, impacts on essential wildlife travel corridors etc., which the proposed mining project would have on the integrity of the World Heritage site, the Federal Government of Canada and the Provincial Government of Alberta subsequently approved the project and published a full EIA in favour of the project. At present the proposed mining project is being legally challenged by conservation groups. IUCN (ed. note: The World Conservation Union) stressed that an increasing number of World Heritage sites (a total of nine, including this case) are threatened by proposed mining projects.

The Committee expressed its serious concerns regarding the impacts that the proposed mining project would have on the integrity of the Canadian Rocky Mountain National Parks and invited the Federal Government of Canada to consult with the Provincial Government of Alberta and to reconsider the decision on the proposed mining project with a view to seeking alternative sites in the region which would have less damaging effects. The committee requested the Canadian authorities to provide detailed information on the proposed mining project, its expected impacts on the World Heritage site, and proposed measures for mitigating those impacts, to the Centre, before 1 May 1998, for review by the Bureau at its next session in mid-1998. The Delegate of Canada indicated that his Government would be happy to provide such a report."[10]

POSITION STATEMENT OF SCIENTISTS ON THE CHEVIOT OPEN-PIT COAL MINE

WHEREAS expert testimony at the joint federal and provincial public hearing on the application to develop an open-pit coal mine at the Mountain Park/Cardinal Divide area concluded that the mine:

- clearly threatens the ecological integrity of Jasper National Park; a World Heritage Site which Canada has committed to protecting the ecological integrity of.

WHEREAS expert testimony pointed out that the mine as proposed would destroy:

- a "hot spot" of biological diversity, including several disjunct, rare and threatened species, as well as a previously unrecorded species for which part of the proposed mine area is to be the type locality for the scientific description of that species;
- a biologically diverse area where there is a high potential for the discovery of additional rare or previously unrecorded species, particularly plants, insects and aquatic life;
- a candidate protected area under Alberta's commitment to complete a network of protected areas and an area recommended by national assessments for designation as a Natural Area of Canadian Significance;
- an area regarded as a refugium that was unglaciated during the last major glaciation and hence, a very uncommon landscape;
- an area rated as an Environmentally Significant Area on a national scale, due to its unique complex of geography, plants and animals of the alpine and sub-alpine of the Rocky Mountains;
- plants from 15 provincially significant and 3 nationally significant populations;
- the habitat of bull trout, which is a species considered threatened in Alberta even though Alberta has yet to officially list threatened fish species;

- an area of song bird species richness and diversity that is "as high as it gets in North America;"
- 4,000 to 5,000 song birds and their off-spring from then on, including birds from 32 species whose populations are already declining in North America; and
- 8 streams which form the headwaters of the McLeod and Cardinal Rivers at the North American continental divide between the Arctic and Atlantic Oceans.

WHEREAS expert testimony also points out that the mine threatens:

- an area already designated as a Critical Wildlife Area "crucial to the maintenance of specific fish and wildlife populations;"
- mammals and birds from populations of 27 species listed as being in trouble in Alberta (Blue, Yellow A and Yellow B listed); and
- a significant population of Harlequin ducks (Yellow A listed), where the only option for the long-term survival of this population is to leave the streams the mine proposes to destroy, untouched and to protect those streams in a natural state.

WHEREAS expert testimony further indicates that for the duration of a century, the mine would result in significant, adverse impacts on:

- important wildlife movement corridors;
- sensitive carnivore species, namely grizzly bears, wolverines, wolves and cougars; and
- prime grizzly bear habitat in a region already slated for efforts to increase the grizzly population as part of the provincial goal to recover numbers to 1000 grizzly bears. And furthermore,
- when the effects of the mine are added to those from the extent of development which already exists or is imminent on adjacent provincial lands, this cumulative impact could lead to the extirpation of grizzly bears, wolves, wolverines and cougars from the region.

and finally, WHEREAS,

- the protection of such an area is precisely what is to be accomplished as part of the Statement of Commitment to Complete Canada's Networks of Protected Areas, and the Canadian Biodiversity Strategy which is Canada's commitment for carrying out the United Nations Convention on Biological Diversity.

WE THE UNDERSIGNED SCIENTISTS call upon the Governments of Alberta and Canada to reject the application for the open- pit coal mine at the Mountain Park/Cardinal Divide location, and to instead apply the principles of the Canadian Biodiversity Strategy and the Statement of Commitment to Complete Canada's Networks of Protected Areas, as well as the findings of the Environmentally Significant Areas assessments by proceeding with formal protection of the entire area of the proposed Cheviot open-pit coal mine.[11]

— THE CHEVIOT JOINT REVIEW PANEL —
REPORT, RECOMMENDATIONS, REACTIONS

The Joint Review panel which examined the Cheviot coal project examined community effects, land use effects, as well as the effects of the project on the aquatic, terrestrial, and atmospheric environments. The following excerpts from the panel's 161 page report sketch out some of the arguments and conclusions which were reached about the project's impact on grizzly bears, Jasper National Park, and the Mountain Park townsite.

I.U.

ON GRIZZLIES AND OTHER CARNIVORES:

Cardinal River Coals:

"CRC noted that carnivores, especially the larger-bodied ones, can be considered both as indicator and as umbrella species for impact assessment purposes. An indicator species in this context is a species that is particularly sensitive to the effects of development and human activities. Measurements of the effects of development on such species provides a measure of the success of impact mitigation programs. For umbrella species, the presence of declines in population and habitat for such species are taken to indicate not only stresses on the species itself, but also on other species and on the ecosystem to which they belong. Protection of the umbrella species, on the other hand, will generally result in the preservation of adequate ecological conditions for other species...[12]

CRC emphasized that it was important to realize that the adverse effects of the Cheviot Coal Project would be part of a regional series of cumulative effects, all of which are already stressing sensitive carnivores. These other sources of cumulative effects included oil and gas activities, quarrying activities, recreational activities, and other mining activities, although some pockets of more secure habitat have been created through access and development control (e.g. designated protected areas).

Overall, the projected regional changes in human developments and activities were predicted to significantly further erode the status of carnivores from an already apparently declining condition... CRC stated that the long-term persistence of populations of grizzly bears, wolves, cougars, wolverines, and fishers, due to ongoing development and activity in the region, was predicted to be threatened within the entire non-national park portion of the study area, and not just the proposed Cheviot mine site...

In their application CRC recognized that the proposed Cheviot Coal Project, along with other regional developments, would have an adverse effect on carnivores. However, CRC also noted that, even without implementation of the Cheviot Coal Project, the currently declining status of the populations and habitats, particularly for large carnivores, is an early indication of the loss of biological diversity in the region.

The mitigation scenarios for carnivores reviewed by CRC included avoidance of disturbance, minimization of disturbance, and compensation for disturbance. Given the scope of the project it was determined that the first two would be difficult, if not impossible to accomplish. CRC stated that the most effective method to address the predicted impacts to carnivores would be to establish a carnivore compensation program. Such a program would be designed to address the needs of carnivores at the regional scale. CRC proposed that, given project approval, it would compensate for unmitigatable losses to carnivore habitat by creation of a "Cheviot Mine Wildlife Compensation Program". In its application CRC committed to: (1) contributing to funding regional research on carnivore ecology; (2) contributing towards establishing and supporting a Wildlife Management Board (or similar body); and (3) contributing toward regional level education packages aimed at informing the public regarding the various carnivores."[13]

The Alberta Wilderness Association Coalition:

"The AWA Coalition stated that, in its view, CRC's evidence confirmed that the best option for carnivores in the region was implementation of a conservation strategy in the region while at the same time denying the Cheviot Coal Project. If the mine were approved, the AWA Coalition indicated that the proposed compensation program should not be seen as an acceptable mitigative measure because: (1) the evidence provided did not show that there is commitment to the compensation program by all stakeholders; (2) even CRC had not made a full commitment to the program; (3) while the program intended for the carnivores to be accommodated outside the Cheviot Coal Project, there is no habitat of comparable quality elsewhere; (4) there was no legal mandate attached to the program; and (5) the lack of experience with such complex collaborative projects made it difficult to predict the chances of success."[14]

The Joint Review Panel:

"In general, the Panel agrees that the available site specific mitigation strategies for carnivores, including (wildlife movement) corridors are, without major and costly changes to CRC's conceptual mine plan, unlikely to be successful in reducing the impacts on carnivore populations significantly. Therefore, the Panel is prepared to consider CRC's proposal to compensate for lost carnivore habitat in areas outside of the Cheviot Coal Project as a reasonable option. The Panel believes that there are now sufficient numbers of models to indicate that such a program would have at least a reasonable probability of success, despite its inherent complexities.

Therefore, should the Cheviot Coal Project be approved, the Panel will require CRC to honour its commitment to act as both a catalyst and a stakeholder in such a process. The Panel will also expect that both the companies and government agencies which advised the Panel at the hearing that they would participate in the program, will also do so. The Panel will require CRC to report to the EUB on an annual basis regarding the status of the program."[15]

ON JASPER NATIONAL PARK

Cardinal River Coals:

"CRC stated that its experience with the Luscar mine revealed that the mine had had little if no effect on the integrity of Jasper National Park. CRC noted the observation at the hearing by Parks Canada officials that there had been no documented problems caused by the existing Luscar mine over the past 27 years...

Based on the information it had gathered, including previous experience with the Luscar mine and the commitment to ongoing communication, CRC did not believe that there would be any significant impacts on Jasper National Park."[16]

Parks Canada:

"As Jasper National Park is contiguous with the proposed project area, Parks Canada was concerned with potential impacts on the Park, which is identified as a world heritage site, from the Cheviot Coal Project. Parks Canada indicated that their primary concern with the proposed Cheviot mine project lies in the fact that any deterioration in the regional ecosystem has the potential to impact Parks Canada's ability to meet its mandate for ecological integrity within the Park...

Parks Canada stated that, in its view, the applicant had insufficiently addressed the issues of cumulative effects. Parks Canada stated that a regional management authority should be established to oversee the development and implementation of strategies to meet the landscape level of goals for the regional ecosystem. Parks Canada stated that this authority should be established by the regulatory agencies of AEP (Alberta Environmental Protection) and Parks Canada and should include major disposition holders in the region."[17]

The Alberta Wilderness Association Coalition:

"The AWA Coalition submitted that Jasper National Park is not an island; rather, it is part of a larger ecosystem. Therefore, based on the policy of the National Parks Act whereby protection of the ecological integrity takes precedence over any other use and the evidence presented at the hearing stating that the region's ecological integrity would be compromised, the AWA Coalition felt that the proposed mine should not be approved."[18]

The Joint Review Panel:

"In assessing the potential impacts of the Cheviot Coal Project on Jasper National Park, the Panel believes it is necessary to be cognizant of both the general east-west orientation of the project and the regional topography. These are highly relevant since, while the western extent of the mine disturbance will be within 3 km of the Park boundary, the vast amount of the areas with high activity (e.g. the coal processing plant) will be several kilometres further away. Furthermore, the presence of significant topographical features between the mine and the Park will further serve to create a barrier between the two land uses. The Panel believes that, overall, CRC's commitments regarding access to the Park, plus its other programs (e.g. the Carnivore Compensation Program), are adequate to mitigate any short-term negative impacts and ensure that longer-term environmental effects are addressed. The Panel does not believe that the Cheviot Coal Project, either on its own or cumulatively, compromises the ecological integrity of Jasper National Park.

With regard to Parks Canada's suggestion that a regional management authority be created, while such a concept may have merit, any such decision would clearly need to be made by both the appropriate provincial and federal authorities when and how they saw fit, undoubtedly following considerably more discussion than occurred at the hearing."[19]

MOUNTAIN PARK TOWNSITE

Cardinal River Coals:

"At the hearing, CRC advised that from the outset of the Cheviot Coal Project, the company had committed to avoid disturbing the Mountain Park cemetery. In order to do so, the company indicated it planned a minimum 60 m buffer between the cemetery boundary and any mine disturbance... CRC noted that it had undertaken a number of modifications to its mine planning and site selection process in order to avoid disturbing as much of the former townsite as possible, while still recovering the important coal reserves beneath the former townsite.

CRC stated that it was prepared to minimize external waste dumping within the area of the old townsite, at significant additional cost, and that protecting the heritage values of the townsite and cemetery were also important factors in its selection of the site for the coal preparation plant and offices."[20]

Mountain Park Environmental Protection and Heritage Association:

"While the Mountain Park Association stated that they were not opposed to coal development per se, they did believe that the Cheviot Coal Project, as designed, would result in unacceptable impacts to the values of the former Mountain Park townsite specifically, and the surrounding area generally. . .

With regard to the cemetery, the Mountain Park Association stated that they were not convinced that CRC's plans to protect the Mountain Park cemetery from disturbance were adequate, and proposed that the buffer between the cemetery and mine disturbance be increased...

With regard to the former townsite, the Mountain Park Association stated that CRC should, at an absolute minimum, be required to leave the entire townsite undisturbed, rather than just a portion. Ideally, the Mountain Park Association stated that it believed that a much greater reduction of surface disturbance in the area around the former townsite was also appropriate. They noted that much of the historical and sentimental value of the site was tied to the surrounding vista and natural beauty of the area. In their view, CRC's plans to create extensive rock drains and external waste rock dumps would destroy these values, as well as destroying irreplaceable habitat for fish and wildlife."[21]

The Joint Review Panel:

"The Panel believes it can understand the concerns raised by the Mountain Park Association, and can sympathize with their feeling of loss. However, the Panel also understands CRC's position that very significant coal resources are buried below the Mountain Park townsite.

In this case, the Panel does not believe that the public interest would be best served by leaving those coal resources in place in order to preserve the entire former townsite... The Panel would urge that the Alberta government place a PNT (Protective Notation) around the area of the former townsite to be preserved in order to provide some additional assurance to the members of the Mountain Park Association that the former townsite has legislated protection. The Panel expects CRC to keep its commitment to work with both Alberta Culture and the Mountain Park Association...

With regard to the Mountain Park cemetery, the Panel notes that AEP (Alberta Environmental Protection) appears to have already used a PNT to create a legislated protected zone around the cemetery. The Panel is not convinced, however, that a 60 m buffer between the cemetery and mine disturbance is adequate. Therefore, the Panel will require CRC to work with staff from the EUB, AEP, and the members of the Mountain Park Association to re-examine the mine plan and establish whether a more substantive buffer can be established. If a larger buffer is feasible, the Panel would urge the Alberta government to expand the PNT appropriately."[22]

RECOMMENDATIONS OF THE JOINT REVIEW PANEL

"The Panel has concluded that sufficient information was provided for it to be able to determine that the majority of the environmental effects, including socio-economic effects, are either positive or where adverse, are not significant. Where the environmental effects were considered to be adverse and significant, they were generally considered to be justified in the context of the project as a whole. In two cases, for the loss of stream (fish) habitat and the loss of carnivore habitat, compensation for non mitigable effects was found to be acceptable.

In one case, that is in upper Prospect Creek, the impacts of mine development were considered to be adverse, significant, and not justifiable given the circumstances. However, the Panel, under the authority granted by the mandate of the EUB, has required that this area be excluded from the Cheviot Coal Project, and so the risk of potential adverse environmentals effects has been addressed. (sic)

Based on this, the Panel recommends that the Cheviot Coal Project receive regulatory approval from the Government of Canada."[23]

REACTIONS

ECONOMY KEY TO CHEVIOT APPROVAL[24]

- by The Edmonton Journal

The key conclusion behind the federal-provincial review panel's approval of the huge Cheviot open-pit coal mine was that the project "will provide significant economic and social benefits, particularly in the Town of Hinton."

The economic arguments in favour of the $250-million Cheviot mine were and remain compelling, and ultimately superseded the environmental consequences, which were judged to be significant and lasting.

It means 450 jobs in Hinton will be preserved once the Luscar mine is closed. Construction will provide up to 900 additional temporary jobs. The mine will generate royalties of $122 million over 20 years.

It also means that the project will result in permanent loss of fish habitat, have a direct impact on water quality both within and beyond mine permit boundaries, and could seriously harm sensitive species such as grizzly bears. This, near the boundary of Jasper National Park.

In the end, there could be no satisfactory compromise, no way that economic and environmental concerns could be reconciled.

The project calls for 32 open pits to pockmark a 23-km-long stretch of the scenic foothills of the Rocky Mountains. Some will be as large as Hawrelak Park and as deep as the height of the High Level Bridge. Little wonder the panel concluded there would be a significant impact on "soil landscapes (and) general terrain features."

How is it possible to "mitigate" against such an extreme and pervasive impact? Dress it up however you want, restock "pit lakes" with fish, but even with the proposals of the company and some additional requirements by the panel – such as to provide a buffer strip along the Cardinal Divide

Natural Area – this is and remains an open pit coal mine.

The landscape that is mined, and some of it has been mined before, will never be the same. You cannot reconstruct pristine wilderness. You either approve of such a mine because of its economic benefits, or you oppose it because of its environmental impact. You cannot have it both ways.

In the end it is the economic argument that won out.

Critics of the project were quick to label the review process a mockery. In fact, the review process was balanced and open to public scrutiny. It involved both the federal and provincial governments. Public hearings were held and input from supporters and critics alike was heard and carefully weighed.

People may disagree with the panel's decision. They may, as one environmentalist did, label it as "despicable." They may threaten as others have done to appeal the decision in the courts, or to launch a high-profile international campaign to try and prevent the mine from being built. But that does not alter the fact that the process which led to the decision was legitimate.

So, too, was the panel's conclusion that to allow the project to proceed "will provide significant economic benefits to both the region and to the province."

ENVIRONMENTALISTS TURN TO THE COURTS

In the wake of the joint review panel's approval of the Cheviot mine a number of environmental groups promised to take the issue to the courts. These promises earned the following indictment from Hinton's weekly newspaper:

I.U.

NO MORE TIME FOR EARTH LOVERS [25]

- by The Hinton Parklander

The news that a powerful environmental group has formed to continue the fight against the startup of this area's Cheviot coal mine project is frustrating. For months Hinton and area residents – Western Canadians for that matter – waited patiently through the public forum process to hear whether or not this nation's coal mining experts would be able to help this region continue to function successfully. We were glad. We were tired. We were celebrating.

But now? Now this. Another attempt by those who claim to love the Earth more than their neighbors to get the support of Albertans.

Environmentalists often come across to everyday folks as being the in-depth, intellectual thinkers who are here for the good of all mankind. It's hogwash. Hug another tree! Ride your bike to work.

Perhaps we're being a bit harsh. But certainly Alberta has been shaped and molded around the success of industrial activity. Environmental concerns have become much more centre stage in recent years, but the overall intent of those who support projects such as the Cheviot coal mine has never been – and never will be – to damage the ground on which we walk, or the air in which we breath. (sic)

Several environmental groups brought up interesting concerns during the panel debate. In fact, the coal mine's environmental experts were forced to change some of their plans because of the knowledge and information brought forward by these special interest groups.

But we must now move on. Some good points have been made. And Cheviot has proven it listens to these concerns – and will continue to.

REBUFFED IN THE FEDERAL COURT OF CANADA

After the federal government approved the Cheviot coal project on October 2, 1997 environmentalists made good on their promise to use the courts in a last ditch effort to prevent the mine project from proceeding. In April 1998, the Sierra Legal Defence Fund argued in the Federal Court of Canada that the joint review panel erred in approving the project. According to the environmentalists, the panel should not have approved the project since the cumulative effects of the project on grizzly bears and other species were unknown and insufficient consideration had been given to underground mining – a less environmentally-damaging alternative. On June 12, 1998 Federal Court Justice William McKeown ruled that the legal challenge was misdirected. Instead of challenging the propriety of the environmental assessment, the judge ruled that the environmental groups should have challenged the federal cabinet's approval of the project. On June 30, 1998 the coalition of environmental groups announced that they were appealing Justice McKeown's decision.

Barring a successful appeal to the Federal Court of Appeal, construction of the mine should begin in the fall of 1998.

I.U.

C H A P T E R 6

BANFF
DEBATING THE LIMITS TO GROWTH

"One of the biggest threats facing the parks is that of development which has been encouraged with increased tourism."

\- World Conservation Monitoring Centre

"What is really at stake in Canada's national parks is our ability to access and enjoy them."

\- Brad J. Pierce,
Association for Mountain Parks Protection and Enjoyment

"If there is a more exquisite pleasure than driving into Banff, it must be watching it recede in one's rear-view mirror... Mammon has set up stall all the way from Bow River almost to the foot of Mount Rundle, a hundred gift shops dispensing life's identical duty-free necessities obtainable at any international airport: cashmere, crystal, Cartier, Caleche."

\- Michael Watkins, *The Sunday Times*, April 30, 1995

— INTRODUCTION —

Contemporary circumstances in national parks around the world tell us a great deal about the nature of modern ecological challenges. They remind us that today's challenges are not posed just by the stereotypical belching smokestacks of industrial society. They arise as well from the public's increased attraction and access to wild places. Leisure, recreation, and tourism have been democratized – changing work/vacation patterns, more well-paid workers, cheap transportation, and the rising importance of leisure are among the factors leading more and more people, from more and more nations, to venture into Canada's national parks. We come not just to see their landscapes but to touch them – to hike them, to ski them, to raft them, to mountain bike them.

Public use of national parks has increased at a staggering clip. To use Banff National Park as an example, the visitors to Banff rose from 500,000 in 1950 to over 5,000,000 in 1995. This phenomenal increase in use hovers over today's debate in Canada's mountain parks between preservationists and populists. For preservationists, we are loving our parks to death. They are likely to nod in agreement with the World Conservation Monitoring Centre's observations. Populists believe preservationists overreact. While it is important to maintain the environmental health of the parks the preservationists' efforts to ban further development are too extreme.

After Jon Whyte introduces us to the textures of Banff, this chapter focuses on the debate over commercial development in Banff National Park. Bill Corbett's article provides an excellent context for approaching today's controversies. Published in 1994, the article sketches arguments which have become perennial features in the debate over whether there is too much development in the mountain parks. From there we move to summarize, all too briefly, the Banff-Bow Valley Task Force Report, an exhaustive look at the state of Banff National Park which called for a radical shift in the orientation of mountain park management. Today's debates are being waged largely within the boundaries of National Park communities like Banff, Lake Louise, and Jasper. The commercial development sought by the Town of Banff served as one of the lightning rods. Our discussion of the tortuous evolution of Banff's plan to increase the size of the town's commercial district highlights the equivocal stance the federal government has taken towards limiting growth. How important are events within the Town's boundaries to the environmental health of the surrounding Park? Two editorials offer competing views on the question. Finally, federal ambivalence towards limiting growth figures prominently in the chapter's concluding section – a look at how the art of "no-growth growth" is being practised at the Château Lake Louise, one of the Park's landmarks.

I.U.

WHEN PEAKS AND PEOPLE MEET[1]

- by Jon Whyte

A major reward for living here is that just when the mountains start looking their best, the light becomes most wonderful, when people who have never put brush to paintpot begin considering a career as painters, and bright odours begin charging the atmosphere, nearly everyone who lacks a reason to be here goes home, and the place becomes nature's again. Nature's and ours.

Numerous photographs by Byron Harmon and many oil sketches by Belmore Browne and Carl Rungius celebrate this most magical of the year's times, the great festival of light which finally burns out like a fuzzy wick in the post-Thanksgiving storms. Those guys went out in the fall as often as in the summer and held on to the wilderness they loved until the threat of winter made them scurry and rush for home, taunting and teasing the season to see how close the first snaps of winter could come to their heels as they skedaddled down from the passes to the relative security of the valleys. A more daunting game when it could be four or five days travel to the highway or the railway.

The cloying scent of valerian, the acrid pungence of slowly rotting sedges, the oblique light and sharp shadow and resilience in our own muscles: from now until the creeks begin to freeze and the leaves have mostly fallen is the festival of the senses.

It's so easy to love the Rockies in July and August that the simplest of emotional types can believe he is a genius of response. As the days become tougher, and shaking off the chill of morning takes longer and longer, as the shrinking days remind of what it means to be Canadian, and autumn begins to make us pull our shoulders in a little tighter, our world becomes more rewarding for the time we spend in it.

In less than two months the air will have lost most of its heat, we shall strive to seek brighter areas for we know that only direct sunlight can keep us momentarily warm, and we'll scurry through the shadows.

Why then as the days grow shorter and dimmer and we know that the promise of bleak November shall certainly be made, why then does the world apparently become so much brighter? Brighter in the foliage, brighter in the contrast of the sunlit slopes and the deep mauves of the shadowed valleys of the mountains, brighter in the light that reflects from the river and through the silver and golden grass, and brighter in the clarity of the sharp scents that are subtler perfumes than the sexy scents of the bee and moth enticing flowers that so profligately wafted their way into the air as recently as a week ago.

Maybe it has something to do with quietness. As the motor homes and the tour buses retreat with their un-Canadian folk – and some Canadians in danger of losing their citizenship because they can't take it anymore! – the place becomes again the kind of quiet world we really want it to be. That is why it seems more correct, more rich, more private, more singlemindedly right than it does in the tumultuous periods of the year when the highway rumbles louder than the waterfall.

Perhaps it had to do with Bart Robinson's observation of twenty years ago when he first saw the wonderful light of autumn in the Canadian Rockies. "Never was light noisier," Bart later recalled. Did it have to do with the serenity and gentle quietness and all that terrifying oblique bright light?

THE BATTLE FOR BANFF:
Environmentalists and Developers Take Off the Gloves[2]

- by Bill Corbett

The turning point, for Harvey Locke, was a walk along the Bow River below the town of Banff in 1987. He found raw faeces and toilet paper floating down the river.

"That was the day I stopped defending Banff National Park and the way it was managed," says the Calgary lawyer and president of the Canadian Parks and Wilderness Society (CPAWS). The town of Banff eventually corrected the sewage problem when it expanded its sewage treatment plant. But the incident convinced Locke that urban and business growth within the park was out of control.

In 1993, Locke and CPAWS launched an international campaign to halt all further development in Banff National Park.

"Banff is a World Heritage Site, it's the second oldest national park in the world and it's a world-famous symbol of wilderness," says Locke. "Yet Banff is the most heavily developed national park in North America by leaps and bounds. And it's being destroyed from within by industrial tourism."

Since 1980, Parks Canada and the Town of Banff have issued more than $500 million in development permits within the park. Locke says the town of Banff has turned into a shopping experience with a mountain backdrop. And, with expansions proposed for golf courses, hotels, housing projects, ski hills and highways within the park, Locke says there's no sign that development is slowing down. He says only a permanent moratorium on development will protect the park from further destruction.

Banff businesses firmly oppose the CPAWS position. In a letter to Banff-area business last may, the owner of the Lake Louise ski resort, Charlie Locke (no relation to Harvey Locke) warned: "As a member of the 'silent majority,' you must speak up now or your parks will be kidnapped by a small but vocal group of publicly funded environmental zealots and extremists."

Three-and-a-half million people visit Banff National Park every year. Banff tourism operators employ thousands of workers and generate hundreds of millions of dollars in annual revenue. They say they already operate within strict development controls that limit expansions to existing leases. Ban all development, they warn, and many responsible companies will go out of business.

"Canadian Pacific has been part of this park since its inception," says Ted Kissane, general manager of the Banff Springs Hotel and regional vice-president of CP Hotels and Resorts. "Because of that relationship and history, we have tremendous investments in the parks – billions of dollars in assets. If we don't maintain and enhance the products we have, the future won't be there."

Last May, the newly formed Association for Mountain Park Protection and Enjoyment (AMPPE) entered the Banff development debate. AMPPE has 1,200 members and ties to the business community in Banff. It calls itself a voice of reason in this increasingly hostile debate, and says it represents the majority of Canadians who want a balance between environmental protection and traditional park uses such as downhill skiing and horseback riding.

AMPPE president and former World Cup downhill skier Ken Reid (sic) says: "We're not pro-development or pro-environment. We're pro-park. Too much of the debate is in the context of all or nothing."

The Calgary Chamber of Commerce, an interested onlooker in the Banff development debate, takes a similar measured position. "We're not in favour of no development," says chamber tourism committee chairman Bill Watson. "There's still room for responsible, sustainable development.

Development has to be environmentally friendly."

Parks Canada finds itself caught in the crossfire as it grapples with a new five-year management plan for the mountain national parks. "We've got a mandate to protect the ecological integrity of the park while maintaining opportunities for people to visit the park," says Banff park planner Judy Otton. "The big question is how that mandate is interpreted."

What Are National Parks For?

Groups like CPAWS argue that the role of national parks is to preserve and protect natural ecosystems. They say protection does not shut down low-key tourism: ordinary visitors can still experience the park in a natural setting. But Harvey Locke draws the line at what he calls industrial tourism: the building of tennis courts, golf courses, bowling alleys and other facilities designed to attract the international convention market.

Banff businesses counter they have been an integral part of the park since its creation and have a right to be there. "The decision was made for us to be in the park years ago," says Kissane. "That's what makes our park unique. Other parks don't have towns or a national highway in them." Park businesses are environmentally responsible, says Kissane, noting his Banff Springs Hotel has eliminated its incinerator and embarked on a major recycling program. But he says they must be allowed to redevelop to compete in today's global tourism business.

"Without that, the Banff Springs Hotel would eventually go bankrupt, because the customer profile has changed," he says. "If we just depended on the Canadian traveller, we wouldn't have the Banff Springs you know today. You have to attract a higher level of clientele to support this infrastructure. It costs $3.4 million a month to run this hotel."

Similarly, Kissane says, the three ski resorts within Banff National Park – Sunshine Village, Lake Louise and Mystic Ridge-Norquay – must modernize to stay in business. "If you look at the ski product in North America, those areas that are viable are the Aspens, Vails and Whistlers. The smaller ski areas are going by the wayside," he says. "It's a relatively flat market and skiers are flocking to the areas with full facilities – accommodation, daycare, restaurants and good lift capacity."

In response, Harvey Locke says promoters of industrial tourism and recreation don't have a divine right to keep growing within a national park. Indeed, he feels development wit-hin (sic) Banff has already greatly exceeded acceptable limits.

The Heart of the Park

No one would call Banff National Park a pristine wilderness. Yet it is far from paved over. Parks Canada has zoned some 85 per cent of the park's 6,640 square kilometres as wilderness. Towns and tourism developments have altered only two per cent of the total landscape. But the location of that two per cent has sparked controversy.

The Bow Valley cuts a narrow path as it winds through the Rocky Mountains within Banff National Park. But the valley contains virtually all the park's montane ecosystem, a threatened habitat both within the mountain national parks and elsewhere in Alberta. Chinook winds keep the montane forests and grassy clearings free of snow much of the winter, providing critical foraging habitat for elk, deer, bighorn sheep and some moose. The Bow Valley also provides an important travel corridor for ungulates, coyotes, wolves, black and grizzly bears and cougars. As such, its health affects wildlife throughout the park.

Ironically, the relatively flat landscape of the valley has proved ideal for development. Within the park, the four-kilometre wide Bow Valley contains the Trans-Canada Highway and the 1A Highway, a national railway, an airstrip, a 27-hole golf course, three ski resorts, the village of Lake Louise and the town of Banff, population 7,500.

The Trans-Canada Highway, a particular sore point in the development debate, acts as both a barrier and deadly passageway for some wildlife species, despite the construction of fences and underpasses along twinned sections. While elk use the underpasses, wolves and grizzly will not. Black bears often climb the fences and cross the highway, and coyotes pursue mice in tall grasses along the road.

"There are more coyotes killed within the park than outside it," says

Harvey Locke. "The highway is an ecological disaster." Before any further twinning goes through, he says, we must find solutions to these problems.

Is the Ecology of the Park Under Threat?

Environmental groups, and some biologists and parks staff, argue all this development in the Bow Valley fragments the montane ecosystem, altering or preventing migration and placing tremendous stresses on wildlife. Radio collar studies, for example, rarely find wolves east of the town of Banff, which acts as a bottleneck to their movements. Not surprisingly the elk wolves prey on hang around the town, causing overcrowding and an increasing number of elk-human confrontations.

"We're putting a large amount of development in the heart of the system," says Locke. "It has horrendous consequences for wildlife, which a national park is supposed to protect."

But others call the threat to wildlife exaggerated. AMPPE president Ken Read notes wolves have made a comeback in the Bow Valley since the 1950s. He acknowledges a decline in moose populations, but says fire suppression and liver fluke parasites probably played a role. "It has been put in a very simplistic context, that the intrusion of man is driving animals away," says Read. "It's much more complex than to say there's too much development."

Environmentalists challenge AMPPE's assertions. "The only data I'm aware of say bears, wolves and cougars are in trouble," says Mike McIvor, vice-president of the Bow Valley Naturalists. "They (AMPPE) say that's science fiction. Let's see their science. The sooner they are forced to produce their data, the better."

A History of Development

The Bow Valley's complexities reflect the tangled history of Banff National Park. The park has a legacy of tourism and development reaching back more than a century. It became a park because CP railway surveyors discovered Banff's now-famous hot springs in 1885. Not long after, CP built the Banff Springs Hotel and the Chateau Lake Louise to entice tourists to the mountain wilderness.

Growth proceeded sporadically until the 1950s, when postwar prosperity and improved highways attracted an influx of middle-class North Americans during the summer. The ski boom of the late 1960s turned Banff into a year-round resort. More tourists meant more hotels and more shops, which in turn meant more housing. But a public outcry killed a proposed major expansion of the Lake Louise ski resort in the early 1970s – the first real organized opposition to development in the park.

Development took off in the 1980s and carried on into the 1990s. The construction boom was most noticeable in the town of Banff, where shopping malls and housing projects sprang up and hotels and motels underwent major facelifts.

Ted Kissane says 90 per cent of this construction has been redevelopment – improvements to land already developed. But Harvey Locke calls it major expansion. He says developers have transformed gas stations in the town of Banff into three-storey shopping malls, and added convention facilities and hundreds of bedrooms to hotels such as the Banff Springs, Chateau Lake Louise and the Rimrock in Banff.

The Debate Heats Up

Calls for a halt to commercial development in Banff National Park are not new. "We've held that position for at least 15 years," says Mike McIvor of the Bow Valley Naturalists. "The first thing that has to happen is development has to stop, and we may well have to head in the other direction." But the campaign to stop development has taken on a new style and size. No longer content to fight development project by project, CPAWS decided in 1993 to intensify the debate.

"It was clear we had to draw a line in the sand and say no more development in the park," says Harvey Locke. Using its national stature, CPAWS has mounted an aggressive public campaign and gained international exposure in publications such as the *Christian Science Monitor. Backpacker* magazine carried a story with the headline "Canada's Flagship Park is Sinking."

Locke says many people agree with CPAWS' opposition to development in Banff. A 1994 study of North America's national parks by the U.S.-based

Brookings Institute calls Banff "unique in its excess." And Locke points to a 1993 Angus Reid survey as evidence of public support for halting development in Canada's national parks.

The survey indicated a strong majority of the 1,365 respondents felt preservation and protection to be the most important priority for Parks Canada. Two-thirds of those who had visited the four mountain national parks felt the area was at or near full capacity, and nearly half thought development in Banff townsite was too high. Only four per cent of Canadians surveyed saw a need for more golf courses or airstrips within national parks.

In late 1993, CPAWS escalated its campaign by launching, in conjunction with the Sierra Legal Defence Fund, a lawsuit against Parks Canada. The suit challenged the authority of Parks Canada to allow Sunshine Village to clear trees for more ski runs without a full environmental assessment first. Sunshine in turn sued Parks Canada, saying the $13-million expansion had already been approved. In October, the Federal Court of Canada ruled against CPAWS and allowed Sunshine to proceed with clearing the runs. But CPAWS claimed a measure of victory since the legal battle prompted Parks Canada to order an environmental assessment of related expansion proposals.

"It's completely unprecedented at the building stage for a project to be halted by an assessment. As far as we're concerned, all the environmental assessments required by law have been completed," says Ralph Scurfield, president of Sunshine Village, which has been operating in the park for some 60 years. He says the project is part of a larger expansion on Goat's Eye Mountain first approved in 1978 and entirely within the resort's leased area. "While I sympathize with the view that the parks shouldn't be paved over, there are a number of existing longterm leases, whether they are ski areas or bungalow camps. Those are legal commitments."

Tourism operators say they can live with strict development controls, tough environmental regulations and extensive public hearings. But they are frustrated with regulatory rules that keep changing.

Judy Otton sympathizes: "The whole idea of environmental assessment legislation is in flux, and I know it's causing commercial operators a lot of grief."

The Job of Solomon

Players on both sides of the development issue mistrust Parks Canada.

In mid-1994, federal Heritage Minister Michel Dupuy handed the job of Solomon to an arm's-length group, the Bow Valley Task Force. The task force has two years to devise a blueprint for ecology-based management and sustainable tourism in the park's Bow Valley. The objective of the blueprint is to forge consensus on the future of Banff National Park.

In early 1995, the task force plans to issue a "state of the valley report": a factual, non-judgemental document outlining such things as wildlife levels, ecosystem stresses, economic activities and park management procedures.

"If we can establish a factual basis right off the bat, then I believe we can get some consensus," says task force chairman Bob Page, dean of the Faculty of Environmental Design at the University of Calgary. "The worst aspect of this is the rhetoric. People have been calling each other liars and questioning motives."

"While the task force deliberates, Parks Canada has declared a moratorium on further development in the Bow Valley within the park. This moratorium excludes several major projects already under review, including further twinning of the Trans-Canada Highway and expansions at Sunshine Village and the Banff Springs golf course. The task force, however, can issue interim reports on the potential impacts of any such projects approved during the next two years.

"This is very much a qualified moratorium" says Mike McIvor of the Bow Valley Naturalists. "I'm waiting to be convinced that it's anything but development as usual."

On the surface, pro- and anti-development forces seem to share some common ground. Both sides say they want to protect the wilderness values that make Banff a park worth visiting. Both sides say development must ultimately face a limit. But environmentalists feel development has already overwhelmed that limit; Banff park businesses believe that limit lies somewhere in the future.

No one knows this minefield of ideological differences better than Page, who is taking a six-month leave of absence from the U of C to work full time on the Bow Valley Task Force. "I'm enough of a realist," he says, "to know that if we misplay our hand, it could easily blow up in our faces."

— DOING SOLOMON'S JOB —
The Banff-Bow Valley Task Force

"This report is a clarion call that continued uncontrolled growth of development and the number of visitors to the area is unsustainable and must be changed – NOW."

- Banff-Bow Valley Task Force, *Banff-Bow Valley: At the Crossroads*

As Bill Corbett points out, in 1994 the job of Solomon was given to the Banff-Bow Valley Task Force, chaired by Bob Page of the University of Calgary. Banff-Bow Valley: At the Crossroads, the final report of the Task Force, answered unequivocally the question posed above: "Is the ecology of the park under threat?" The Task Force concluded that Banff's popularity with the world's tourists, the pressures the park faced from population growth in the central Rockies, the continuing national importance of the existing transportation network through the Rockies, and cutbacks in Parks Canada's budget meant that: "The ecological integrity of the Banff-Bow Valley cannot be sustained."[3] Animals the AMPPE thought were recovering nicely were, according to the Task Force, faring poorly. Banff's status as a national park was threatened.

The number of recommendations made by the Task Force – more than 500 – underlined the depth of the problems plaguing Canada's oldest national park and the unacceptability of the status quo. Some recommendations to reduce animal-human conflicts – like fencing the Banff townsite and outlying hotels and resorts – were too much for Parks Canada to accept. Others, controversial though they proved to be, were embraced quickly by Sheila Copps, the Canadian Heritage Minister: no new land for commercial development in the Park, closing the Banff airstrip and bison paddock, relocating the cadet camp and horse corrals.

The Task Force preached "aggressive growth management," a euphemism for very little or no commercial growth. It also called for governments to licence and favour enterprises which would further the Banff-Bow Valley view of what Banff's primary goal should be – restoring the Park's ecological integrity. Task Force members called for Parks Canada, the Town of Banff, and the Park's business community to adopt a "Touchstone Tourism Destination" model. The philosophy of this model differed dramatically from what many of Banff's businesses practised. The model emphasized "the theme of learning, education, understanding and appreciation of nature and the Rocky Mountain (sic)... It will be the "glue" that binds the efforts of all who seek to realize the tourism potential of the Banff-Bow Valley."[4] A good or moral business supplied basic services like food or lodging or services clearly related to the goals of a national park (such as guided natural history tours).

When the Task Force addressed the Town of Banff directly it urged the Town's mayor and councillors to slam the brakes on growth. It asked the Town to accept a future as the prototype of the Tourism Destination model. The Town's growth forecasts should undergo "a fundamental reassessment." Commerce in Banff should be realigned to conform to the premises of the new tourism model. Luxury retail outlets had no place in this vision of the town's future: "Implicit in this model is a shift away from some of the clothing, jewellery, souvenir and other specialty shops that are not directly linked to the heritage values of the Park. These, over time, could be phased out or converted to more appropriate enterprises... "[5] In other words, implement this alternative model aggressively and the likes of Nesbitt Burns, Ralph Lauren, and Jacques Cartier would become endangered species on Banff's streets.

I.U.

— BANFF'S CONTROVERSIAL — COMMUNITY PLAN

"When it comes to running the future and the mechanics of the Town, it's up to the Banff people, not to outsiders."

- Ossie Treutler Sr., Banff businessman and former Town Councillor

When the Banff-Bow Valley Report was released in October 1996 it delivered a strong anti-growth recommendation to the federal government and Banff Town Council. As two events – the development of the Town of Banff's ill-fated Community Plan and a new convention centre at Lake Louise – illustrate, neither the federal ministers responsible for our national parks nor Banff's municipal politicians have the stomach to prohibit new growth in the Park. Between September 1997 and June 1998 Heritage Minister Copps and Banff's Mayor Ted Hart debated the Town's intention to allow its commercial district to grow by 850,000 square feet (one environmental group likened the Town's plan to building a 30 storey office tower or 20 new motels). During some of the debate's less-dignified moments – for example, when Copps accused Hart and the rest of the Town Council of "crass commercialization" – you could be excused for seeing the Minister as ecological integrity's tough talking avenging angel. Yet, as some of the Minister's actions during the Community Plan debate suggest, seeing her in this light exaggerates the strength of her commitment to stop growth in its tracks.

"We want the Town of Banff to remain a Town – a Town that respects its very special place within a National Park. I am asking the Mayor and the Council to ensure that within the community plan, to be completed in 1997, the Town's population remains below 10,000 permanent residents."

- Canadian Heritage Minister Sheila Copps, October 7, 1996.

The Minister's directions to Banff Town Council concerning the Town's plans for future growth indicated that, her rhetoric aside, she too imagined a future where Banff's commercial district would be allowed to expand. When she released the Banff-Bow Valley Report she told the town of 7,600 that the Council's community plan "should feature a permanent population of less than 10,000 residents." One of the quirks of life in a National Park concerns the requirements you must satisfy in order to maintain a permanent residence in a Park. Permanent residents basically must work in the Park, own a business in the Park, retire after working or owning a business in the Park, or be a spouse or dependent of the worker, owner, or retiree.

In other words, Banff could only grow towards the Minister's limit if the size of the business community increased. The amount of new commercial space sought by Town Council was married to Council's desire to get as close to the Minister's cap of 10,000 residents as possible. According to studies commissioned by the Town, 850,000 square feet of new commercial space would boost the municipality's permanent population by 2,323 people and put Banff just an eyelash short of the population growth limit set by Copps. By setting the population parameters she did for the Town of Banff, the Minister invited Council members to indulge their desires to let Banff's commercial district grow.

In September 1997, Sheila Copps flatly rejected the Banff Community Plan, blaming its proposed level of commercial development for her decision. Yet, neither the Council nor the people of Banff were ever told how much new space, if any, she would accept. Instead of clear directions, the Minister served a double-barreled ambiguity: "In total we need a plan that leaves neutral environmental impact. To achieve that, obviously the numbers will probably change."[6] What did neutral (no net) environmental impact mean? Obviously, the numbers will probably change? For the remainder of the fall Mayor Hart insisted that 850,000 feet of new commercial space was still feasible – an opinion that was not denied by either Copps or Andy Mitchell, her Secretary of State for Parks and the politician charged with working with Banff Council on a revised Community Plan. Had the Minister given Council her bottom line on how much future growth was acceptable (presuming, too generously perhaps, that Parks Canada had a bottom line on the issue), months of uncertainty could have been avoided.

Throughout the fall and winter of 1997-98 Banff Council worked with Mitchell to try to satisfy Ottawa's concerns about the original plan. The most difficult issues were: commercial growth rates, the activities of an "appropriate" business in a national park, and the meaning/measurement of the principle of "no net environmental impact." The Mayor persevered in believing that Parks Canada would accept the amount of new commercial space proposed in the June 1997 plan, especially since the Town added the stipulation that the 850,000 square feet would be phased in over fifteen years. The results of a March 26th plebiscite strengthened his resolve to press ahead with the original amount of commercial space. Ironically, the plebiscite question did not allow Banff's residents to comment on whether they thought the 850,000 square feet proposed by Council was appropriate. Instead, they were asked if they favoured more than 850,00 square feet (12% in favour), between 650,000 and 850,000 (43% in favour), less than 650,000 (13% in favour), and absolutely no growth at all (31% in favour). Within two months of the plebiscite victory, Mayor Hart had sent the revised Banff Community Plan back to Copps.

The Heritage Minister unceremoniously dashed Banff's commercial growth hopes when she stopped briefly in Banff for the photo opportunity which comes with opening the Banff Television Festival. Copps, seldom one to mince her words, saw the Council's repackaging of its commercial growth intentions as "unfettered and unabated development" that threatened the World Heritage Status Banff shared with Canada's other Mountain Parks.

She rebuked the Town:

> I think that when you live in a national park and you have the privilege which comes with living in a national park you also have additional responsibilities and those responsibilities include ensuring that in your quest for additional commercial space you don't lose sight of the fact that there's more to life than shopping centres.[7]

Two days later, in the House of Commons, she continued to portray Council as being interested in little more than bringing more shopping opportunities to Banff:

> It is unfortunate what the Banff council did. Instead of seizing an opportunity to create a real ecocommunity into the 21st century, it chose crass commercialization.[8]

These remarks were too much for Hart to swallow. He lashed back that Copps was treating the people of Banff like "pond scum" and that his town was just a pawn in her efforts to make Canada look good on the international environmental stage.[9] On previous occasions Hart argued that Banff was being singled out unfairly. It was being treated as if the activities within its boundaries were the only matters influencing the ecological integrity of Banff National Park. If controlling growth was important, where were the signs that this principle was being applied to Lake Louise? or Jasper in Jasper National Park?

More importantly for the long term future of the Bow Valley, what about Canmore? From 1991 to 1996 Canmore grew by 45 per cent, and had reached a population of 8,396 people. If the town continues to grow at this rate, its population in 2010 will be nearly 20,000 people. By then, the town plans to have added 4,000 hotel rooms. If this type of growth materializes, what type of effect will Canmore have upon the ecological integrity of Banff National Park? Hart's message here was clear and important – regional cooperation is crucial to securing the ecological integrity of Banff National Park.

Judging by the tough talk of the Heritage Minister in the first part of June it looked like Ottawa and the Town of Banff had irreconcilable differences on the issue of commercial development in the townsite. This appearance was deceiving since, by the end of the month, these differences evaporated. Copps announced that, in the coming fall, Ottawa would legislate limits on commercial development in National Park communities. The announcement also addressed the specifics of the commercial expansion proposed by the Banff Community Plan. Ottawa cut the amount of space it would allow from 850,000 to 350,000 square feet. Surprisingly, Hart did not react strongly against the federal fait accompli. Instead, he called the Copps announcement "a win-win situation for everybody."[10]

Why was the Mayor now so accommodating? In addition to cutting back Banff's proposed expansion, Ottawa agreed to purchase the Pinewoods property – 5.5 acres of commercial property Hart called the "most problematic" issue the Town had to deal with since it gained municipal status in 1990. Ottawa's move, an initiative Hart previously had urged Ottawa to take, eliminated the threat of a lawsuit against the Town (the property's owners said they might sue because Banff's intention to spread commercial growth out over fifteen years would prevent them from building the 250 room hotel project they planned). Furthermore, Banff promised to invest millions of dollars in the property by constructing an environmental education centre. The issue of commercial expansion in the town was resolved – at least temporarily.

As environmentalists congratulated the Heritage Minister – and some Banff businesses berated her – several fundamental ecological questions were forgotten: How would the amount of commercial expansion Ottawa committed itself to – 350,000 square feet – improve the ecological integrity of the Park? What information did Ottawa rely upon to conclude that this expansion would not increase the risks to the Park which the Bow Valley Task Force had identified so forcefully less than two years earlier?

I.U.

BANFF IN PERSPECTIVE[11]

- by The Globe and Mail

Banff National Park is one of the world's natural wonders. Environment Minister Sheila Copps has aptly described it as "a living work of God's art, a place of such raw visual strength that it staggers the human mind." If you believe Canada's environmental movement, it is also a fractured ecosystem pushed to the brink of collapse. The main culprit: unchecked commercial development.

Environmentalists argue that development in and around Banff, Alta., the tourism town in the heart of the park, has left the natural habitat degraded and wildlife in danger. They were backed up by an environmental report released last fall. The Banff-Bow Valley Study said that human activity had damaged the area so severely that Banff was in danger of failing to qualify as a national park.

Ms. Copps reacted quickly to the study. She said no new land would be released for commercial use. She closed an airfield, a bison paddock and a military cadet camp that were said to interfere with the movement of animals. That was not enough for some environmentalists. When Banff town council submitted a plan to allow 850,000 square feet of new development on existing land, they protested loudly. So, this week, Ms Copps rejected the plan, saying it would create "a town without a soul."

Let's try to put this dispute in perspective. The Town of Banff occupies about five square kilometres. Banff National Park occupies 6,640 square kilometres. Add in three other mountain parks along the Alberta-British Columbia border – Jasper, Yoho and Kootenay – and you have a continuous stretch of parkland about four times the size of Prince Edward Island.

True, Banff park is heavily used. About five million people visited last year. Most stayed in the area near the town, a sensitive "montane" environment of grassland and forest where elk, wolves and bear come to feed. (In fact many tourists never venture outside the town limits.) But the permanent residents number only 6,000 and Ms. Copps has decreed that Ottawa will never let the population exceed 10,000.

Most townspeople have no problem with that. They only want to improve and expand their facilities to accommodate the tourists, who contribute about $750-million a year to the Canadian economy.

The town's development plan would not increase the size of Banff, whose borders are fixed by a 1989 act of Parliament. All of the development would take place on land already zoned for commercial use. Because there are no unoccupied tracts of commercial land left in Banff, most of the growth would take place through "infilling" putting more square footage on already occupied lots. No major new resorts or hotels were envisioned, because there simply isn't the space. No new building would be more than three storeys high; town rules prohibit it.

Overall, the amount of commercial space in Banff would grow by one-third, much less than the town had planned a few years ago. A 1992 plan, approved by Ottawa, would have allowed for 2.45 million square feet of new space, three times what the town now seeks. Precisely because of the environmental concerns, Banff has scaled back. It has also taken steps to limit its impact on the environment. The town is putting a special wildlife corridor through one of its neighbourhoods. It has built a state-of-the-art sewage treatment system. And it has installed water-conservation devices in all local residences free of charge.

Yet, somehow the impression has formed that Banff is a kind of Rocky Mountain Coney island, a sprawling blot on the natural environment, a future Aspen, Colorado. This is silly. The environment of Banff National Park may in fact be under threat, but it is not because of the town of Banff. The number of elk in the area has risen. The wolf population is coming back. Walk a few minutes outside the busy town centre and you can be in the heart of nature.

Everyone wants to preserve the natural beauty of the park. That includes the people of Banff. Their livelihood, after all, depends on it. But it would be a mistake to turn the park into a museum piece. Banff is known around the world because people come from around the world to see it. Many of them stay or stop in the town of Banff. They deserve decent, modern facilities. As long as those facilities respect the surrounding habitat, Banff should be allowed to build them.

BANFF GROWTH MUST BE HALTED:

Copps's Announcement was Tactless[12]

- by The Edmonton Journal

On Sunday Sheila Copps said she wasn't going to talk about the Banff development plan. On Monday she torpedoed it.

While her decision to reject further commercial expansion of the Banff townsite is a sound one, the tactless way she made the announcement leaves a lot to be desired.

Copps did not have the courtesy to meet with Banff town councillors and tell them why their months of work was rejected. She just blabbed to reporters at the Banff Television Festival.

This issue has been a real struggle for the people of Banff. When the town council's first development plan was rejected, the council went at it again, refined the plan, got approval in a local plebiscite, and finalized it in May before sending it to the minister.

Given that effort, Mayor Ted Hart deserved to hear Copps's decision directly, not through the media.

The town of Banff is an anomaly, a municipality inside a national park. The plan the town council came up with, 850,000 square feet of new commercial space developed over the next 15 years, would make total sense in any other town – a balance between business pressures and environmental concerns.

Inside Canada's most overdeveloped park, it's not acceptable. Banff is in terrible trouble, as the 1996 Banff-Bow Valley Task Force warned so eloquently.

It said: "Five million visitors every year, two communities, a trans-continental railway, a four-lane highway, three major ski hills – this growth in visitor numbers and development threatens the mountain environment. If allowed to continue, it will cause serious and irreversible harm to Banff National Park's ecological integrity and its value as a national park."

Yes, development has to stop. It won't be easy: tourists will keep coming, looking for hotel rooms and restaurant meals. Those hotels and restaurants will end up priced out of reach of ordinary Canadians; that's how supply and demand works.

Yet even if it makes no sense to Banff council, the town of Banff has all the shops and restaurants it should have. Building more will simply encourage more people to come.

Back in 1996, Copps responded to the task force report by promising to curb development. However, her signals since then have been mixed. Just last month her department approved a convention centre and 262 new hotel rooms for Lake Louise.

Those mixed signals have been unfair to the people of Banff, wrestling with their development pressures, and worrisome for all those concerned about the future of our parks.

The minister needs to be very clear about just what will and won't be allowed in the park. If there's to be no further commercial development, then Copps should say so and Banff councillors won't have to waste their time in future.

Such a clear signal may be the legislation setting out a clear blueprint for development in national parks that Copps has promised to introduce in the next few weeks.

That sort of blueprint cannot come too soon, since development pressures on the mountain parks are only going to increase. Capping growth in Banff will prompt developers to seek expansion in other parks – indeed, Jasper Park Lodge is already seeking a 154-room expansion.

The days when Canada had unlimited wilderness are long gone. Only quick and decisive federal action can protect that unique wilderness that is our mountain parks.

— LAKE LOUISE, THE CHÂTEAU — AND "NO-GROWTH GROWTH"

At the height of the controversy over business expansion in Banff Sheila Copps accused the Town Council of bowing to "crass commercialization." Yet, this slur could have as easily been directed at her own office in light of decisions her Ministry was making at the same time for Lake Louise, the hamlet just 55 kilometres up the highway from Banff.

Since the first of Cornelius Van Horne's "fire-wagons" rumbled through the Bow Valley in the 1880s Canadian Pacific has dominated the Canadian psyche and the Valley's economy. Punching the Canadian Pacific Railway's mainline through the Rockies, as told by Lightfoot or Berton, was a mythic act for a young Canada. With the railway came the discovery of Banff's hotsprings and then the destination hotels the railway built to cater to elite tourists from all over the Western world. Few would quarrel with Canadian Pacific's claim that its Banff Springs Hotel and Château Lake Louise "are virtual icons of Canadiana, part of the very fabric of Canada." As concerns grew about the health of Banff, these icons became more controversial. For environmentalists, their continued profitability demands year-round use and occupancy rates which only increase the stress on the Park's wildlife and landscape.

Canadian Pacific Hotels has made some efforts to address the concerns of environmentalists. They contribute to large carnivore research studies; they no longer plan to expand further the Banff Springs golf course; they have reduced the number of rooms at the Banff Springs Hotel. But, in the case of Château Lake Louise, Canadian Pacific also shows signs of the "no-growth growth" syndrome. The syndrome appears when the plans for growth from developers and their political champions are cast in a no-growth light. The no-growth light is generated by arguing that the growth a developer wants is actually below the growth ceiling set in some living or dead management plan. These ceilings, without justification, become surrogates for ecological health. If these established growth limits are exceeded, the Park will suffer. If they are not exceeded, the Park will be fine.

The syndrome appears in all of the debates about whether or not there is enough development in the parks. For example, it was prominent during the battle between Ottawa and Banff Council over the Town's ambitions to grow – 850,000 square feet of new space was really a reduction since a 1990 Parks Canada plan would have allowed twice as much new commercial space as what Banff would have if Ottawa approved the Town's proposal.[13] It's a wonder that the promoters of growth in our national parks don't go back and look to the ambitions of Sir John A. Macdonald or Cornelius Van Horne to justify their endeavours.

"No more growth in either the hamlet of Lake Louise or at the Château Lake Louise" – this is the conclusion a trusting reader of average intelligence could draw from Sheila Copps's reference to Lake Louise in her speech marking the release of the Banff-Bow Valley Task Force Report:

> "As for Lake Louise, the Lake Louise Action Plan will be adopted. The overnight commercial capacity of the Lake Louise Area will remain at 3,500. And the number of rooms at the Château Lake Louise will not exceed the current hotel capacity. And again, I thank CP Hotels for agreeing to that."

These remarks are phrased in the language of "no-growth growth." It would have been just as accurate, if politically riskier, for the Minister to have said the following:

> "As for Lake Louise, the Lake Louise Action Plan will be adopted. The overnight commercial capacity of the Lake Louise Area will remain at 3,500. This is 1,151 – or 45 per cent – more people than the 2,349 overnight visitors Lake Louise's lodgings can currently accommodate. And the

> number of rooms at the Château Lake Louise will not exceed the current hotel capacity. The Château has applied, however, to increase their overnight guest capacity from 994 to 1,126, a thirteen percent increase, but still a little below the overnight visitor cap the Château has agreed to. The Château's growth is under the ceiling of growth set for Lake Louise in the 1997 Banff National Park Management Plan. And again, I thank CP Hotels for agreeing to that."

In May 1998, the government announced that, after a limited federal environmental assessment, the growth sought by the Château had been granted. Canadian Pacific had received permission to incorporate a seven storey convention centre into the hotel. The convention centre included a 700 person meeting hall, eighty-one guest rooms, and a 252 seat dining hall. In order to try to camouflage the significance of the facility Canadian Pacific argued disingenuously that it was "designed to serve registered hotel guests, rather than attract outside visitors." The size of the Château's expansion was nearly half as large as the expansion Ottawa would force Banff to accept six weeks later; the Château's expansion would come all at once – Banff's would be phased in over time.

Some steps were taken to try to temper the uproar the government and Canadian Pacific knew would be coming from environmentalists. Canadian Pacific agreed to return 20.5 acres of undeveloped lease land to the Park and to restore another 22 acres of its lease with the area's natural vegetation. Ottawa lowered Lake Louise's overnight capacity ceiling from 3,500 visitors to 3,100 (but, this figure was still thirty per cent higher than the maximum number of overnight visitors who could use Lake Louise now). As has too often been the practice in Canada's mountain parks, the federal government also promised – after the fact of development – to develop ecological benchmarks and indicators and to monitor and research them to ensure that, in the long term, there would not be any net negative environmental impact. In other words, the real environmental assessment of the convention centre would not begin until after Canadian Pacific had completed its expansion.

Environmental and wildlife scientists have expressed their concerns that the Château's plans will worsen the plight of threatened species such as the Park's grizzlies. "More overnight visitors at Lake Louise," warned Stephen Herrero, "will likely contribute further stress to an already stressed habitat and population."[14] Suzanne Bayley, one of the Banff-Bow Valley Task Force members, was very discouraged by the federal government's decision. Referring to the Task Force's work she said: "We recommended that they should control growth and in fact stop growth. The park will be in dire trouble both ecologically and socially if it continues to go this way."[15]

These concerns did not move the federal government. The June 1998 announcement that Ottawa would legislate ceilings on commercial development in National Park communities included a declaration that there would be a moratorium on commercial development in the communities administered by Parks Canada. The Château's convention centre was exempted from this moratorium. One might be forgiven for thinking that, when it comes to calling the shots in the Bow Valley, not much has changed since the 1880s.

I.U.

CHAPTER 7

THE TIMBER BEAST
AN ENDANGERED SPECIES?

The lumber king, the timber beast,
Is on the rampage all the time,
Slaughtering the hardwood,
The poplar, spruce and pine.

He is the despoiler of our North woods,
Who lays waste God's handiwork,
A person who despoils your heritage,
Is one, who all men should shirk.

- from "The Timber Beast"
by Henry Stelfox[1]

— INTRODUCTION —

In the past decade, Alberta's forests have grown rapidly in their economic and political importance. The Alberta Forest Products Association claims that the industry is now the province's third largest (following energy and agriculture and ahead of tourism) and employs approximately 40,000 people. In this chapter we look at some of the controversies which have accompanied the industry's growth in the foothills. The first set of controversies revolves around the plans of Sunpine Forest Products Limited to increase the timber harvest from the central foothills southwest of Rocky Mountain House. Sunpine's plans raise questions about the federal government's commitment to its environmental assessment process and the commitment of Sunpine and the Government of Alberta to practice ecological management – a progressive style of forest management. From there we proceed to consider two of the issues – clearcutting and biodiversity – which often are presented in ways that tarnish the environmental image of the forest industry. In this section we offer two accounts that challenge the public's perceptions that clearcutting is inherently destructive and that industrial forestry is necessarily harmful to biodiversity. The chapter concludes with a detailed look at the Chinchaga – a section of the northern foothills that, although identified as a prime candidate for protection from development, seems instead destined to be opened up to extensive timber harvesting operations. I.U.

— MAINLINING THROUGH — THE WEST COUNTRY

Booming – that's one word to describe the West Country – the portion of the foothills laying between Rocky Mountain House and Sundre in the east to the front ranges of the Rockies in the west. In the past ten years, the West Country has become a hotbed of industrial activity. Seismic crews and exploratory wells probe deep beneath the forests, farmlands, and muskeg for petroleum pools earlier, less sophisticated, searches may have missed. Where these explorations succeed – and many have – new pipelines follow, criss-crossing the countryside and waterways, to get product to processing facilities.

As the number of logging trucks rolling down the region's main and secondary roads suggests, the forest products industry is booming here too. For decades now forestry has been an important feature of the West Country's socio-economic life. Its historic significance contributed to Rocky's selection as Canada's Forest Capital in 1961, its rediscovered significance since the late 1980s contributed to the town's selection as Alberta's Forest Capital in 1996.

As in other regions of Alberta, the logging boom has generated opposition. In the Rocky Mountain House area the opposition has been led by Friends of the West Country, a local grassroots group. Like many of the forestry-oriented groups that joined the Alberta environmental movement in the late 1980s, the Friends organized when increased appetites for timber threatened their lifestyle. Since 1989 much of the group's attention has been focused upon Sunpine Forest Products, the most significant forest products player in the regional economy. Depending on the yardstick used, the company employs anywhere between 320 and 800 people. The company claims to have invested $110 million in its Sundre and Strachan facilities since 1987.

In 1992, the company's importance to the future of the forests in the central foothills grew when it was awarded forest management privileges over 2,312 square miles. Soon after receiving these privileges, Sunpine announced plans to carve 41 kilometres of all-weather logging road – the Main Line road – through the heart of the West Country. For the Friends and, as we'll soon see, for some provincial officials too, the Main Line road was an environmental folly. During its 41 kilometre length it would cross 21 rivers and streams – some of Alberta's finest trout habitat. By opening up the wilderness to all terrain vehicles, the new road would put unwelcome pressure on the region's fish and wildlife resources. Moreover, citizens and experts alike argued that the road simply wasn't needed. Existing all-weather roads – notably the North Fork road – although not as convenient for Sunpine and rural residents – could be used to penetrate the interior of the company's forest management area.

The battle over the Main Line road exposes several environmental controversies. One of these centres on the issue of public involvement in assessing the environmental impacts of forest industry operations. What informational requirements have to be satisfied for the public to play a meaningful role in environmental assessments?

A second goes to the scope of the environmental assessments required by federal environmental assessment legislation. Is it enough for the federal authorities to consider just the environmental impact which will result from the bridges the Main Line road requires? Should they consider the environmental impact of other activities associated with the bridges such as the road itself? the forestry operations that provide the rationale for the construction of the bridges?

Finally, the Main Line road controversy touches the commitment of government and industry to "walk the talk" of respecting other resource values, in this case fish and wildlife values, which are essential to alternative economic pursuits such as tourism. I.U.

PUBLIC INVOLVEMENT IN ENVIRONMENTAL ASSESSMENTS

At first glance it's hard not to be impressed by the commitments governments make to public involvement in environmental policy making. Democratizing policy making, opening it up to the general public and groups which historically have not been invited to play the policy game, has become fashionable. Respect for public participation often figures prominently when governments announce their latest policy fashion. For example, when Alberta's Minister of Environmental Protection unveiled "The Alberta Forest Legacy" – his ministry's version of sustainable forest management – his press release claimed that the Forest Legacy built upon the recommendations that "were prepared with broad public input and stakeholder involvement."[2] Elsewhere, Canada's Commissioner of the Environment and Sustainable Development notes: "Public participation in the environmental assessment process is one of the fundamental principles set out in the Act."[3]

Companies like the view from the public participation bandwagon too. Sunpine Forest Products has committed itself to a partnership approach to forest management. "Public input," the company promises, "plays an important role in Sunpine's proposed Detailed Forest Management Plan. Sunpine cannot be a responsible steward of the forest resources without it."

The reality is arguably less wholesome. In the past two years the federal government's environmental assessment process has been rebuked from several quarters, perhaps most notably by investigative journalist Andrew Nikiforuk and Brian Emmett, Canada's Commissioner of the Environment and Sustainable Development. Nikiforuk's stinging attack on the state of environmental assessment opens by characterizing environmental assessment as "cynical, irrational and highly discretionary". In part, he uses such damning language because the public cannot get the information needed to make a meaningful contribution to environmental assessment proceedings. The lawyers, business leaders, and citizens he interviewed agreed unanimously that the Canadian Environmental Assessment Act, at more than 50 pages in length, was far too complicated. As Ian Scott of the Canadian Association of Petroleum Producers asks: "If you want public input then why does the government make it so complicated that the public can't get into it?"[4]

Bruce Emmett's language, although more restrained, makes the identical point. "Good communication of information," he writes, "is essential in order to obtain public input." Yet, there was little about the public information system associated with federal environmental assessments which impressed him in his May 1998 report. In fact, he concluded: "Having an information source that is neither timely nor complete, that is difficult to access and to use, and that may not meet Agency guidelines compromises the principle of facilitating public participation in the environmental assessment of projects."

These general commentaries will come as no surprise to the Friends of the West Country. They capture well the difficulties which the Friends faced as they tried to question the plans of Sunpine to build the Main Line road. The Department Fisheries and Oceans, a department Nikiforuk concludes to be "the government department most reluctant to apply the act," balked at assessing the environmental impact of the Main Line road on the region's fisheries resources. Once the Department agreed to examine the planned road, it refused to give the Friends the documents they were entitled to, documents vital to their ability to contribute to an environmental assessment of the Sunpine logging road.

Frustrated by the department's failure to comply with the law, the Friends turned to the courts. The Friends' complaint that the federal government was withholding vital information was heard by Justice Muldoon of the Federal Court of Canada. Justice Muldoon, a hero to some environmentalists for his 1989 order that the federal government could not licence the Rafferty-Alameda dams on Saskatchewan's Souris river until it did an environmental assessment, agreed with virtually all of the complaints lodged by the Friends against the federal government. The Federal Court Justice's low opinion of the government's behaviour is enunciated clearly in the following excerpt from his judgment, an excerpt which ends with his conclusion that the Department of Fisherics and Oceans must divulge much more information to the Friends of the West Country than it had originally. He reached similar conclusions in regards to four of the other five information requests made by the Friends of the West Country.

The following brief excerpt refers to Rule 1612 which is a Federal Court rule compelling the respondent (here, the Department of Fisheries and Oceans) to produce documents in the department's possession. Section 55 is a section of the Canadian Environmental Assessment Act which creates a public registry of documents. With limited exceptions, the registry must contain all records produced, collected, or submitted concerning an environmental assessment.

I.U.

The Friends of the West Coast Country Association vs. The Minister of Fisheries and Oceans

"In sum, the CEAA requires that the responsible authority shall maintain a public registry for public access, subject to limited exceptions. The registry is public; not private or even cloistered. The language in section 55 is clear: if there is going to be an environmental assessment, there must be a public registry and that registry shall have all documents relevant to the proposal. (One does wonder if the drafters of the CEAA were vying with those of the Income Tax Act to achieve the greater complexity or to discourage Court challenges.)...

It is, as the applicant (editor's note: Friends of the West Country Association) has pointed out, self evident that the material in question is in the possession of the responsible authority (editor's note: Department of Fisheries and Oceans). The language of section 55 is very broad and inclusive. For example, pursuant to paragraph 55(3)(a) "any report relating to an assessment" must be maintained in the public registry; 55(3)(b) "any comments filed by the public" must be maintained. The inclusivity of this language also brings out another self evident feature of the registry, namely that anything contained in the registry is relevant...

According to Decary J.A., the tribunal does not have to do anything other than to hand over relevant material in their possession. The tribunal does not have to find new evidence for the applicant. In the context of the CEAA process, this means all information in the respondents' possession concerning Sunpine's projects is relevant to the specific proposal...

Request #1

'All documents contained in the public registry regarding the Prairie Creek and Ram River bridge proposals.'

Response

'Any documents contained in the public registry that were considered by the tribunal have been certified and filed on October 21, 1996. In our view this complies with rule 1612.'

The applicant seeks the certified copy of Sunpine's proposal document. The only parts of the proposal document included in the tribunal record were pp. i, 1, 3 and 4 of the proposal document and "several scattered pages" of one of 10 appendices. The actual proposal is 21 pages long and has 10 appendices. (The respondent provided an uncertified copy of the proposal to the applicant on October 22, 1996 (affidavit of Carol McDonald), but apparently there has not yet been any certified copy forwarded to the registry or given to the applicant).

The applicant argues it is entitled to a certified copy of Sunpine's proposal because it provided the basis for the screening report and the NWPA (editor's note: Navigable Waters Protection Act) proposals. Not only is the proposal required to be in the public registry, but is clearly relevant when the grounds for the originating notice of motion are examined. This is at the heart of this judicial review application. The respondents' argument, that only relevant information which was considered by the tribunal needs to be given to the applicant, is not supported by the jurisprudence. It is hard to imagine that the responsible authority considered only 4 pages of a 21 page proposal. Relied upon, perhaps, but reliance is not part of rule 1612. (How were the pages selected without considering the whole document?) Perhaps the respondents are just too wan, busy or tired to comply. Could it be?

Further, the applicant alleges that other documents which were in the public registry were omitted by the respondents. These include comments on the proposal document which were submitted by the Alberta Environmental Protection - Surface Water Rights Branch and the Alberta Environmental Protection - Resources Administration Division, which the screening reports indicate were received. The applicant also seeks any other documents in the public registry of which the respondents are aware. This is not an attempt at discovery: these requests are for documents already in the public eye. These documents are relevant and must be certified and turned over, even if it does seem to take too much tiresome trouble and effort for the respondents."

Editor's Postscript: In September 1997, the Friends of the West Country finally received the records they were seeking from the federal government. From the documents they learned that the federal government essentially failed to consider nineteen of the twenty-one stream crossings.

HOW FAR TO GO?
The Scope of Federal Environmental Assessments

A second issue raised by the controversy over the Main Line Road concerns the scope or the inclusiveness of a federal environmental assessment. What exactly does the federal government have to consider when it examines a project requiring federal approval? In respect to the Main Line Road, since the road crossed Prairie Creek and the Ram River – navigable waters according to federal legislation – those crossings required federal approval. The federal government interpreted its responsibilities to assess the stream crossings very narrowly. It felt bound only to consider the environmental impacts arising from the construction of the bridges and the structures themselves. In Ottawa's opinion, the bridges Sunpine wanted to stretch over these waterways "were not likely to cause significant adverse environmental effects." A comprehensive environmental assessment of the bridge projects was unnecessary. In August 1996 Sunpine received Ottawa's blessing to construct its bridges.

The Friends of the West Country argued in the Federal Court of Canada that the federal government failed to comply with the Canadian Environmental Assessment Act. Ottawa broke its own law. Given what Nikiforuk called the "myopic" view the Canadian judiciary has of the environmental assessment process, British bookmakers probably would have given the Friends long odds on their legal gambit winning. But, strike pay dirt they did. Justice Frederick Gibson agreed that the scope of the federal environmental assessment of the Ram River/Prairie Creek bridges was far too narrow. "I conclude," he wrote, "that the responsible authority was obliged by section 15 (3) to include within the scope of the environmental assessment the road, and, perhaps, the forestry operations, if they were "... in relation to... " the projects, that is to say the bridge and their related abutments. The responsible authority was without discretion."

Justice Gibson reinforced this conclusion when he shifted to consider the implications of the section of the federal legislation (section 16.1) outlining a requirement that the screening of a project must consider the "cumulative environmental effects that are likely to result from the project in combination with other projects or activities that have been or will be carried out." Again, the Federal Court Justice drew the commonsensical conclusion that the bridges had been built to open up the West Country for forestry operations. The implications of this conclusion for the scope of the environmental screening Ottawa must carry out were enormous. Gibson wrote:

> I read subsection 16(1) to require a responsible authority to include within an environmental screening the environmental effects that may occur in connection with the project as defined and also those that are likely to result from the project in combination with other projects or activities that have been or will be carried out. Indeed, the evidence before me indicated that it had already, at least to a very substantial degree, been carried out at the time of the hearing before me. Similarly, the evidence before me would appear to indicate that, with the approvals under subsection 5(1) NWPA, the proponent's proposed forestry operations in the West Country will also be carried out. Once again, I conclude that subsection 16(1) clearly reflects, on the facts of this case, an obligation on the part of the responsible authority to apply the independent utility principle in the definition of the scope of the assessment.

Ottawa, in other words, was obliged to broaden the scope of the assessment to include at least the Main Line road and possibly the forestry operations.

Environmentalists celebrated Gibson's decision. If the ruling stands, they will be one huge step closer to guaranteeing that the federal government brings a comprehensive eye to the environmental assessment process. In Edmonton, the decision was vilified. Ty Lund, never one to miss an opportunity to lash out at federal institutions, called the decision "so ludicrous that it's unbelievable." For Lund, Sunpine had taken every precaution imaginable and the Friends of the West Country should be chastised for believing that the democratic process didn't end whenever Lund's department said it did.[5]

The implications of this decision extend well beyond the forests of the West Country. The environmental coalition fighting the Cheviot mine believes Gibson's generous reading of Canadian Environmental Assessment Act requirements will assist their arguments before the Federal Court of Appeal. Environmentalists in Banff have taken the Château Lake Louise convention centre to court. They claim that the federal assessment of the project was too narrow and that, following Gibson's ruling, the cumulative environmental effects of projects and activities linked to the convention centre must be considered too. Other provincial governments will take note of the decision too since it opens the door to federal intrusions in areas of provincial constitutional jurisdiction.

I.U.

ABOUT ECOLOGICAL MANAGEMENT in the West Country

"These people have a basic premise that thou shalt not cut a tree."

- Bruce Buchanan, President, Sunpine Forest Products

Changes in the range of values and interests which foresters incorporate into decision-making is one gauge of forest management's sensitivity to environmentalism. Contemporary environmental debates have pushed forest managers to begin to think seriously about forests as ecological systems, not just as fibre factories. Today, forest companies and governments are just as likely to argue that non-timber resource values should be central to resource use decisions as they are to proclaim their enthusiasm for public involvement. The debate over the construction of the Main Line road is one piece of evidence regarding the genuineness of the commitment Sunpine Forest Products and the Minister of Environmental Protection have made to elevate the importance of non-timber values. The debate over this road to resources shows how difficult it may be to put ecological management into practice. It requires measures of political and corporate will which could not be found in the West Country's forests and streams in the 1990s. We now compare the statements made by Sunpine and the Ministry of the Environment about the importance of ecological management to the advice the Ministry of Environmental Protection received – and rejected – on the ecological impact of the Main Line road.

The Commitments

The commitments of Sunpine and the Alberta government to respect non-timber values are made in various documents and statements. Here, we rely upon excerpts from "The Alberta Forest Legacy", Alberta's implementation framework for sustainable forest management as well as Sunpine's overview of the detailed forest management plan it prepared in 1996.

I.U.

Alberta and Ecological Management

"Ecological management is the practice of designing our activities on forest lands so that they do not interfere with the ecosystem's ability to perpetuate itself, as it has for millennia. In some instances this might lead to limits on consumption; in others it will increase economic and societal opportunity within the forest. The Provincial Government will play a role in developing and promoting ecological management as a tool. Citizens and industry will use this form of management as a tool as they plan and assess activities according to the impact of those activities on the ecosystem. Ecological management applies to the full range of land-use scales, from localized stand-level planning to broad landscape-level planning. It emphasizes the contribution of science and public input in forming hypotheses and framing desired outcomes. It leads us to plan our activities so that they maintain, to the extent possible, the level of inherent disturbance processes common to any particular ecosystem, so that the diversity and productivity of the forest landscape is maintained."[6]

Sunpine and Ecological Management

Sunpine's attitude towards ecological management is taken from a section of the forest management plan overview entitled "Objectives of Forest Management Plan." Three of the company's objectives are summarized here – ecological management, integrating resource uses, and forest management. The other objectives mentioned in the plan are access management, operations, forest protection, inventory/research/modeling, public involvement/education, and timber supply. The company writes:

"Ecological Management

Ecological management is an evolving approach to forest management that focuses on ecological processes and ecosystem health, while sustaining benefits that people derive from the forest.[7]

1. Maintain the long-term health of the forest by managing for a diversity of habitats, species and genetic resources...
2. Recognize that the FMA is one component of a large, regional landscape and must be managed in a manner compatible and consistent with overall regional plans...

3. Continue development of ecological management practices that are appropriate for the FMA area...

Integrating Resource Uses

Although its management responsibility is for the timber resource, Sunpine recognizes that its management practices will impact on other resource values such as wildlife and fisheries. Integrating resource users is an important component of Sunpine's ecological management strategy.

1. Manage the timber resources according to objectives and guidelines described in the Sub-regional Integrated Resource Plans and other resource plans approved by the province for the FMA area...
2. Incorporate the needs of the many users of the FMA area and mitigate potential conflicts...
3. Identify and protect or minimize impacts to sites of historic, environmental, ecological, recreational or cultural significance...
4. Cooperate with other agencies in the enhancement of non-timber resource values, including tourism and recreation...

Forest Management

Sunpine's forest management practices are intended to provide the company with a sustained yield of timber for its processing facilities.

1. Conduct forest practices which maintain a sustained yield of timber from the productive forest land while adhering to the principles of ecological forest management. The productivity of the land will not be diminished; other resource values will be considered in all phases of planning and operations...."[8]

FORGOTTEN VOICES
Government Criticisms of the Main Line Road

"This is the first documented evidence we've had that he (Environmental Protection Minister Ty Lund) totally ignored environmental protection in favor of development."

- Martha Kostuch, 7 February 1996

One of the noteworthy aspects of the Sunpine controversy is a paper trail – memoranda, letters, and meeting notes – we rarely have the opportunity to follow in most conflicts between environmentalists, forestry companies, and government. In the conflict over the Main Line road the Friends of the West Country used Alberta's freedom of information legislation to secure advisory opinions authored by the government's stewards of non-timber resources such as fish and wildlife. In the following excerpts from the documents the Friends obtained, government officials raise serious objections about the ecological impact of the road Sunpine blazed through the West Country.

I.U.

December 9, 1993. Memorandum from L.A. Rhude, Fisheries Biologist to George Robertson, Head, Timber Management, Rocky Clearwater Forest

"The proposed road would cut through an area which up to now is relatively inaccessible. Improve[d] access would result in increased use of the area. This would negatively impact the fisheries by increasing harvest.

The building of a new road would increase the amount of runoff resulting in increased discharge. During storm events the amount of water entering the creeks would increase resulting in increased natural erosion. Spring runoff would also be larger than currently experienced.

Considerable instream work would be required at all river crossings.

Instead of building a new haul road, I would recommend that Sunpine use the North Fork Road. The North Fork Road will probably have to be upgraded resulting in silt entering the creeks during the construction phase and up until the new ditches are revegetated. This will occur at either site.

On the plus side, there would be less environmental impact by using the North Fork Road: a) no new country would be opened up, and b) runoff would not be significantly increased over current levels. The North Fork Road crosses fewer streams and lower down in their watersheds than the proposed haul road resulting in less impact."

December 14, 1993. Memorandum from Brian Burrington, Wildlife Technician, to George Robertson, Head, Timber Management, Rocky Clearwater Forest

"I see no reason to duplicate existing road access as proposed.

The proposed access road will negatively impact wildlife populations and watercourses throughout its entire length. Secluded headwaters access on numerous streams will be subjected to heavy long term environmental damage as well as nonsustainable pressure on the fisheries and wildlife by various user groups. I strongly recommend that the section proposed from 3-38-9-W5 to 2-39-12-W5 be deleted and existing access by utilized." (sic)

July 6, 1995. Memorandum from Robert Glover, Chief Ranger, Clearwater District, to Lorne Goff, Regional Director, Eastern Southern Slopes Region

"Surely, one road, one bridge on the Ram, shared and used jointly is a better alternative than Sunpine and the MD maintaining TWO separate roads and bridges.

The North Fork road is being assessed as if it were a new construction proposal in terms of disturbance, creek crossings, etc. Realistically, this environmental impact of upgrading, while significant, cannot be equated with totally new construction.

The fragmentation of the habitat into smaller areas and more accessible (by road or ATV) is a MAJOR concern from the wildlife point of view.

The NFR (Editor's note: North Fork Road) is assessed as new construction. The impact has already occurred on the ecosystem and will continue even if a mainline road is built. Constructing the mainline simply adds to the ecosystems impacted."

August 22, 1995. Letter from Brian Burrington to George Robertson

"There doesn't appear to be any new information or answers concerning the questions about wildlife or fisheries in the supplement. There are no suggested mitigative measures included for either resource.

Fragmentation of habitat and wildlands is still a major concern.

Contrary to Sunpine's statement there is no way to construct a road across a watercourse without habitat loss and some long term impact to the stream and/or fishery."

August 25, 1995. Letter from Lorne Goff, Alberta Forest Service, to Keith Branter, Sunpine Forest Products Ltd.

"The preferred Main Line Route corridor as proposed in Sunpine's Revised Phase 1 submission and Supplementary Information, and as conditionally endorsed by the Forest Advisory Committee is approved by the Department.

The approval of the Mainline Route is based on consideration of social, economic and engineering factors as detailed in the proposal and supplementary information, recognizing that the route may impact wildlife, fisheries, and water quality, and that mitigation measures will be required by the company during construction and operation of the road."

— CLEARCUTTING — "NOT A DIRTY WORD"

"There is a perception that clearcutting is a primitive forestry practice. This is not necessarily the case... It is often more appropriate than any other method for... species such as lodgepole pine and aspen. While the public perception of clearcutting is of massive harvest areas, the practice includes cuts as small as a fraction of a hectare as well as units of hundreds of hectares."

- Alberta, Expert Review Panel on Forest Management

Of all the things that foresters do," writes Hamish Kimmins, "clearcutting probably causes the most outrage and anger in the public."[9] Albertans are likely to agree with him. When Alberta's Expert Review Panel on Forest Management asked Albertans what concerns they had about forestry practices, clearcutting figured prominently in their responses. The public has been receptive to the environmentalists' message that clearcutting = deforestation. Clearcutting, this message insists, has turned Canada into "The Brazil of the North." When Macmillan Bloedel, one of Canada's largest forest products companies, announced it was going to phase out clearcutting in the old growth forests of coastal British Columbia the company won kudos from environmental organizations. Are the public's concerns justified? Have environmentalists presented a distorted picture of clearcutting's environmental impact?

I.U.

CLEAR-CUTTING:

Industry expert addresses clear-cutting issues[10]

- by Wayne Thorp

"... There are many public concerns and conflicting opinions about clear-cutting in Alberta's northern forests. As a professional forester, I believe that myths and misconceptions about clear-cutting are the cause, and in the following I will attempt to present the facts objectively and show that clear-cutting is not a dirty word.

Misconception #1

Clear-cutting creates vast areas of total deforestation.

Technically, clear-cutting means the removal of all merchantable trees within a patch of specified size within a forested area, and the size varies with the species of tree and other factors...

(Thorp pointed out that, depending on the species, the average size of the clearcuts or patch cuts made by Daishowa were between 40 and 98 acres.)

These patches are also restricted in shape and location out of regard for other resource values such as stream side buffers, aesthetics and wildlife habitat considerations such as sight distance, travel corridors and mineral licks.

The number of patches that can be cut within a given forested area similarly is restricted. As a general rule no more than 50 per cent of the merchantable stands within a planning area can be removed in the first pass. Subsequent removal of the reserved stands cannot be done until the previously cut areas are fully reforested and provide adequate hiding cover for wildlife...

Misconception #2

Clear-cutting is an unsuitable method of harvesting if the aim is sustained yield of the forest.

In most areas of the world, excepting only the perpetually wet tropical rain forests, our forests have been burned over at more or less frequent intervals for many thousands of years. Recognition of this is particularly important in northern Alberta where over 80 per cent of the area has had a forest fire in the last 80 years.

Therefore, although we might like to think so, what Mother nature (sic) provides us now is not a first growth forest, but rather a sustained yield forest. Most of it has been reforested naturally after having been clear-cut in nature's way by fire.

There are lessons for us in the result. For instance, the species we have in northern Alberta are the ones that recognize certain sites through natural succession. They would not occur in their current composition or distribution without this fire history.

Recently, forest managers have become increasingly aware that we must strive to understand the reasons for the success Mother Nature has had in re-establishing forests following such disturbances as fire, and to apply this knowledge in forest management that involves clear-cutting yet aims at sustained yield. To do this it is essential to treat each species individually, recognizing its special needs.

In terms of forest biology, a clear-cut results in a dramatic change in conditions which may make regeneration easy or difficult depending on the species. To begin with, the climate of the site has changed in terms of light, temperature, humidity and wind. The variation in extreme values between day and night and between different seasons of the year becomes greater.

There is also a change in water and nutrient conditions. The water level in the ground rises and there may be increases in run-off water.

In order to select appropriate forest management practices we must look at the origin and growth requirements of the natural forest and the associated tree species within it based on the local climate, soils and topography.

Lodgepole and Jack Pine:

Lodgepole Pine and Jack Pine are short-lived small to medium-sized coniferous threes that generally grow on drier sites than spruce.

The species are shade intolerant (do not re-establish or grow well under shade), and have serotinous seed bearing cones which require fire or other disturbance to open and release seeds. The cones once formed on the tree will retain their seed for periods up to 15 years before opening unless they are subjected to enough heat to break the bond on the cone scales.

This difficulty combined with the fact that germination of seeds and establishment of the seedlings in shade would be an unsuitable environment, selective cutting would do more harm than good as a management system for pine.

Foresters have learned that they can create the same conditions as fire for pine by clear-cut logging a patch, which causes the cones to break off the tree and lie on the ground. Site preparation usually with crawler tractors mixing the moss and soil creates a suitable seed bed for germination.

The final component needed to open the serotinous cones is heat. Complete removal of the canopy allows sunlight to raise surface temperatures to the 47°C threshold necessary for opening the cones (sic) scales.

As can be seen in this example, clear-cutting is an imitation of the forest fires that Mother Nature used to sustain this species. But clear-cutting has the advantage that forest managers, with pubic (sic) input, have complete control over the size, shape and location of the patch. Moreover there is none of the damage done (to human settlements and to wildlife) when vast areas were burned instead.

Of course, artificial reforestation by tree planting also speeds up the process of re-establishing new trees and the achievement of sustainable yield... "

— BIODIVERSITY — AND THE TIMBER INDUSTRY

Anyone who accuses the timber industry of deforestation is likely also to accuse forestry companies of destroying the ecosystem's biodiversity – the number and variety of organisms living in a forested ecosystem. Those who object to clearcutting as a method of harvesting trees in the foothills are likely to believe that clearcutting always damages an ecosystem's biodiversity. Many professional foresters feel this perception is too simplistic. Sure, clearcutting, like many other types of human activity over the ages, may reduce the biodiversity of a forest. But, the disturbances of clearcutting may increase biodiversity too. In other words, the relationship between biodiversity and clearcutting is more complex than the critics of this forest practice would suggest.

As part of its contribution to the debate over appropriate forestry practices, the Alberta Forest Products Association publishes an educational resources series. The material below is excerpted from the sixth issue of this series.

The excerpts refer to alpha, beta, and temporal diversity. Alpha diversity refers to the species living in a particular ecosystem (such as the foothills). Beta diversity refers to the species diversity you will encounter in portions of a particular ecosystem (such as in a valley or a low marshy portion of the foothills). Temporal diversity refers to the changes in alpha and beta diversity over time.

I.U.

BIODIVERSITY

An Environmental Imperative[11]

- by Dr. J. P. (Hamish) Kimmins

"Concerns about biodiversity

There are several reasons why people should be concerned about conserving biodiversity. However, sometimes the concept of a species right to live is confused with the right of every organism to live. The latter idea is clearly in conflict with our understanding of evolution, natural selection and ecology. Nature always produces more offspring than can possibly survive. When more survive than was intended, the "balance of nature" is upset. The idea of keeping all species from going extinct is also in conflict with our understanding of the process of evolution. Species are going extinct all the time through a variety of natural causes, and always will. It is a question of maintaining a balance between species extinctions and the evolution of new species, as well as maintaining the species on which we put a particular value.

Maintaining biodiversity to enable future generations to enjoy our personal experiences of nature also reflects the antipathy of people towards change.

It is said that we should conserve tropical forests out of self-interest if for no other reason. There is undoubtedly a reservoir of undescribed and unresearched life forms in the forest that will be of future medical and commercial benefit to humans. These benefits will only be assured if we conserve the biodiversity of the forests long enough to discover and research these life forms.

"The basic rule of intelligent tinkering is to keep all the parts," said American conservationist Aldo Leopold. This advice has sometimes been interpreted to mean that all ecosystems should have the same species list all the time. This is not what Leopold meant. This wrong interpretation of a good idea overlooks the fact that ecosystems always go through a wide range of alpha species and structural diversity following natural or human-caused disturbances. A particular ecosystem will experience a wide variation in the makeup of the microbial, animal and plant community over time, yet function normally throughout the period of redevelopment following disturbance.

The problem of species loss is that in some ecosystems there may be one or more individual species that play critical roles. Without this role, ecosystem function will be impaired, but we may not know which of the species in the ecosystems plays this essential role.

We should attempt to "keep all the parts" within historical range of variation in an ecosystem. A logical strategy would be to examine the alpha, beta and temporal diversity of a local landscape and record the range of diversities in the local ecosystems, and ensure that within this range, all the ecosystem components are kept across the forest landscape.

Biodiversity goals must be defined

Many people believe biodiversity is a key issue that must be considered in forest management. Some feel that the only way to conserve biodiversity is to preserve large areas of unmanaged "wilderness" forest. Others feel that by modifying management practices, both alpha and beta diversity can be maintained while continuing to harvest some reduced level of commodities from the landscape.

The major difficulty in the biodiversity debate is that biodiversity goals have not adequately been defined. Unless these goals are defined, people may force polices (sic) that achieve goals in terms of one aspect of biodiversity, while failing in terms of another.

Some people fear that if forest management continues, major species losses will occur and many ecosystems will be destroyed forever. However, we must recognize that much of our forest biodiversity (geographic, regional, and some aspects of beta diversity) reflects different climates, soils, geologies and topographies, and many of the species depend on periodic ecosystem disturbances. Biodiversity experts know of very few cases of species extinction in Canadian forests that have been caused by forestry. Although there has undoubtedly been changes in some aspects of biodiversity as a result of forest management.

Some people also believe that the alpha, and some aspects of beta diversity found today is the way nature "intended" it to be. Present diversity levels reflect the past history, natural and human-caused disturbance, as well as present conditions. There does not appear to be any particular "right" or "wrong" level of biodiversity, or any biodiversity "master plan" that nature has.

Disturbance is a part of natural forest ecosystems. Many species have adapted to disturbance and will disappear in the absence of disturbance. There is a need to understand the response of species and aspects of bio-diversity to the various types, levels, patterns and frequencies of disturbance. This way forest management systems can be designed to permit the conservation of social values of forestry, as well as the environmental values of biodiversity.

Effects of timber management on biodiversity

Clearcutting – Where clearcutting is conducted as scattered patches in a previously unbroken expanse of even-aged forest, it can increase beta diversity, both species and structural. On any particular site it may reduce structural diversity by removing the tree layer but increase plant species diversity as a diverse early plant community develops. The effect on animal diversity will depend on the effect of the harvesting on wildlife habitat values. Animal species diversity may increase, decrease, or remain unchanged even though there is a change in species.

Continuous clearcutting of large areas may have little effect, or may reduce beta diversity. This will depend on the pre-harvest landscape diversity. Clearcutting will generally create forests of the same age and increase the overall variety of plant species initially, while reducing the diversity of certain plant groups. These effects do not persist. There will be a progressive change in both structural and species diversity over time following the harvest.

Alternatives to clearcutting can have significantly different effects.

- Selection harvesting (Editor's note: the periodic harvesting of large trees – either individually or in small groups) will generally maintain or increase alpha structural diversity but may have little effect on alpha species diversity and on beta diversity, or may reduce it.
- Shelterwood harvesting (Editor's note: a less dramatic type of clear cutting, a heavy thinning of the trees) will maintain structural diversity better than clearcutting for as long as the residual trees are retained..."

— CHINCHAGA —

Give With One Hand, Take Back With the Other

Few members of the general public have heard of the Chinchaga Caribou Range – approximately 6,000 square kilometres of northern foothills drained by the Chinchaga River. This area, centred about 220 kilometres northwest of the town of Peace River, is more familiar to members or observers of Alberta's energy and forest products industries. Looking at maps of Forest Management Agreement (FMA) areas, of logging on public lands, and of oil/gas wellsites scattered across my walls and floor it's apparent why resource companies are drawn to the Chinchaga. By the standards set elsewhere in the foothills, the Chinchaga is relatively unexplored, relatively untouched. The density of wellsites, the levels of seismic and logging activity, are considerably lower in the Chinchaga than the levels found to the north, south, and east of the area. (For example, a 1992 aerial photograph of a township just north of Chinchaga showed that this 93 square kilometres hosted more than 180 cutlines and dozens of wellsites.) More importantly, significant oil and gas discoveries such as Chevron Canada Resources' 1995 wells into the Muskeg formation – one-and-a-half miles below the surface – confirm the commercial potential of the region.

The Chinchaga's relatively undisturbed nature also has attracted the attention of provincial and national environmental organizations. Since the Chinchaga is not as encumbered by industrial commitments and activities as most other places in the foothills organizations like the Canadian Parks and Wilderness Society (CPAWS) and the World Wildlife Fund (WWF), regard it as a good candidate for protection from extensive industrial use. Here, some divisions of the provincial government agree. A provincial assessment of protected areas candidates in the foothills concluded that the Chinchaga's "comparatively 'undisturbed' character" makes the area "by far the best candidate for the location of a major Foothills wilderness area." This character, plus the range's environmental significance, led the Protected Areas Division to designate the Chinchaga as one of the thirteen best candidates for protection in the foothills.

This designation belies the fate of the Chinchaga. The Minister of Environmental Protection has approved Forest Service recommendations that make the Chinchaga a "best candidate" for more industrialization, not for protection. In September 1997, the Land and Forest Service added nearly 450 square miles of forests to the timber licence of Manning Diversified Forest Products Ltd. This addition included about one-half of the territory which a provincial advisory committee had recommended should be set aside as the Chinchaga Special Place. Ty Lund, the Minister of Environmental Protection, knew of this recommendation; he approved this recommendation. By rubber stamping the Land and Forest Service decision, Lund also gave the green light to new logging in this candidate for Special Places status.

Environmentalists were outraged by the change in the Manning timber licence. They felt, with justification according to the minutes of the Special Places Provincial Coordinating Committee, that there was a consensus among stakeholders on what the boundaries of the Chinchaga should be. The permission to cut spruce from these new lands also seems clearly to violate the spirit of a provincial policy (interim protection measures) designed to insure that areas under consideration by the Special Places process should not be opened to new development initiatives. For logging in the foothills, the policy allowed planned timber operations to continue. But, the Land and Forest Service was not supposed to approve new timber licences.[12]

Why did the Minister take back with one hand what he had given with the other? It might be plausible to suggest that compromising the Chinchaga was accidental. The original boundaries for this Special Places candidate did not include the territory added to Manning's timber licence. Since this additional territory was not included in the original candidate site, Lands and Forests, caught by surprise, assumed it was business as usual – get the cut off the land. Had they only known that the Chinchaga boundaries might be moved it would never have amended Manning's licence. If accurate, this view points to serious sclerosis and poor communication within the Department of Environmental Protection. The department is represented on the provincial Special Places committee. The committee had been examining the possibility of moving the Chinchaga boundaries for some time (as the minutes clearly show). Lands and Forests should have known that the committee was moving the boundaries and changed its licence amendment intentions accordingly.

For some environmentalists, this interpretation of events is far too charitable. They see a government determined to give resource extraction primacy over the protection of natural landscapes and processes. The government's refusal to value environmentally significant areas highly enough is the source of controversies like the Chinchaga. This refusal is seen as the rule, not the exception. This conclusion led the rather moderate Canadian Parks and Wilderness Society (CPAWS) to resign from the provincial Special Places committee. For CPAWS, the government's commitment to protection, the primary goal of the Special Places program, is as empty as a pauper's pockets. "We need to see action from this government," argued Wendy Francis, "not more words. Their words are hollow when it comes to environmental protection and can't be believed."[13]

I.U.

WHY THE CHINCHAGA IS A CANDIDATE
for the Special Places Program

In 1996, the Department of Environmental Protection released the report Selecting Protected Areas: The Foothills Natural Region of Alberta. The following excerpt highlights why the Chinchaga was selected as one of the best candidate sites in the foothills. The candidate site ultimately selected is much smaller than the area described below (roughly one-sixth the size). The site did not include lands in the Daishowa-Marubeni FMA or Reserve Areas. The acronym ESA stands for environmentally significant area.

I.U.

"2.2.12 Chinchaga Caribou Range (ESA 1702)

ASSETS: "Large, key Woodland Caribou range; estimated population of Chinchaga herd was 200 animals in 1991; includes a large diversity of landforms and vegetation communities including mature and old-growth forest." Also includes "important Grizzly Bear habitat along the Chinchaga River" and "a number of Trumpeter Swan nesting lakes south of the Chinchaga River." Contains "a large diversity of wetland types including patterned and non-patterned fens, bogs, marshes and lakes." No population centres within ESA. Appears to be free of coal dispositions.

PROBLEMS: To the north and east, roughly 40% of the ESA falls within the Daishowa-Marubeni Reserve and one township lies in their FMA area. The ESA wholly overlies 11 gas fields, parts of two others and portions of two oil fields. Contains roads, gas plants, airstrips, pipelines, transmission lines, seismic cutlines and trails; however, for the most part, degree of fragmentation is not as severe as rest of Foothills Natural Region. ESA contains roughly 0-12 wellsites per township, and has a low-to-moderate cutline density of 20 - 40 per township (mostly 1992 data). Area immediately to north of ESA has been intensely fragmented as a result of oil and gas industry activity.

COMMENTS: The large area is poorly known from a biophysical standpoint. As drawn, the ESA boundary outlines the range of the Chinchaga caribou herd, but should be amended to include Halverson Ridge, the ca. 65 x 10 km, NE - SW oriented linear tract of Upper Foothills terrain adjacent to its southeastern margin. The size, quality, diversity and comparatively 'undisturbed' character of ESA 1702 make it by far the best candidate for the location of a major Foothills wilderness area. From a biodiversity conservation/protected areas design perspective, the two highest priority, potential protected areas within this ESA are: (a) 'Lakeland' i.e., the southern, lake-rich... portion extended to encompass part or (preferably) all of Halverson Ridge and, (b) the north-central portion (including part of Tanghe Creek) with the (Upper Foothills) Fontas Hills area at its core. This ESA constitutes the Foothills' best opportunity for a genuine, large protected area."

A CORPORATE VIEW
of the Need for a Chinchaga Protected Area

Daishowa-Marubeni International Ltd. (DMI) is the forest products company with the most extensive Forest Management Agreement area in the Chinchaga area. In an October 2, 1996 letter, Bob Wynes, DMI's Forest Resource Coordinator explains why the company felt that a large, protected area should be established. No company held an FMA in the timber management units (P7, P8) mentioned in the letter. These timber management units have been, however, proposed as part of the FMA which would be given to Grande Alberta Paper, a company proposing to build a paper plant near Grande Prairie. DMI's suggestion that the candidate site be located north of the Chinchaga River was not followed; virtually all of the candidate site recommended by the Provincial Coordinating Committee lay south of the Chinchaga River in P7. The acronym AAC stands for Allowable Annual Cut – an amount of timber to be harvested annually which is roughly equal to the amount of new growth produced by the forest each year.

This letter is also important for what it says about the process for selecting Special Places candidate sites. The Provincial Coordinating Committee (PCC) referred to in the letter is supposed to play a key role in identifying candidate sites. Regarding the Chinchaga, an interdepartmental committee of officials from the provincial departments of Environmental Protection/Energy, Economic Development and Tourism/Agriculture, Food and Rural Development/ Community Development/ and Municipal Affairs usurped this responsibility. Before the Chinchaga nomination reached the PCC, the interdepartmental committee reduced the size of the candidate site from 7,000 to 1,000 km^2. I.U.

"We recently received your referral for the Chinchaga Special Places 2000 candidate site. I would like to provide the following information for you and the Provincial Coordinating Committee to consider.

Daishowa-Marubeni International Ltd. (DMI) is developing a new approach to Detailed Forest Management planning which is based on more closely emulating the natural patterns and processes of the forest. The intent it to develop an approach to forest management which has a high probability of maintaining biodiversity, and a sustainable supply of wood fibre. One of the key components of this approach is recognizing the need to maintain some ecological benchmarks across northwestern Alberta to provide the opportunity to compare the results of our future forest management practices with the natural (forest fire origin) forests. Ideally, these benchmark areas would be as large as feasible to provide the best opportunity for natural processes to continue. The forests in the benchmark areas need to be representative of the forests which are being harvested. We feel that representative ecological benchmarks are one of the primary values of protected areas.

DMI supports the development of a network of protected areas. This network should be developed with several objectives in mind... If we only address the SP2000 objectives now, the future demands to meet the other objectives will result in a greater cumulative impact on industry, and likely not meet the objectives as well as they could have been. In addition to the need for ecological benchmarks described above, we need to consider other concerns such as caribou and wilderness values.

Specifically regarding the Chinchaga candidate site, DMI feels that a 1000-2000 km^2 protected area is possible and appropriate within the boundary of the original Canadian Parks and Wilderness Society (CPAWS) nomination. We do not however, feel that the best area within this boundary was selected by the government inter-departmental committee. The area south of the Chinchaga indicated in the referral is largely peatland and relatively homogeneous. It contains only the Lower Foothills natural region and is not very representative of much of the Foothills area in northwestern Alberta. We feel that it has limited value as a representative ecological benchmark for the foothills. We anticipate that this would result in the need to expand this area in the future to include better representation, which would result in a greater cumulative impact on industry than selecting the appropriate site now.

An alternate boundary for a site north of the Chinchaga River, within the original boundary of the area nominated by The Canadian Parks and Wilderness Society (CPAWS), is indicated on the attached map. (Editor's note: The alternate area suggested by DMI covered about 18 townships in P8; 45 townships/partial townships are found in P8.) We feel this area is more appropriate for the Chinchaga SP2000 site as it provides a much greater diversity of landscapes, including both Upper and Lower Foothills natural regions. It contains a mixture of peatland and mineral soils and a diversity of forest stand types. It is quite representative of the mixture of sites found in the foothills in northwestern Alberta and would be much more valuable as an ecological benchmark. The area we propose north of the Chinchaga would be a reduction of approximately 60% of the land base of P8, or 43,800 m^3 of deciduous AAC. The impact of the candidate site south of the Chinchaga would be a reduction of approximately 31% land base of

P7, or about 34,900 m^3 deciduous AAC. These estimates are based on the assumption that there is an even distribution of stands in these management units. DMI has deciduous quota in both P7 and P8.

In terms of other resource values, the attached map also indicates information obtained from the North West Region Standing Committee for Caribou. The data was collected by the Natural Resources Service between 1991 and 1994 by tracking radio collared caribou. During the winters of 1992/93 and 1993/94... caribou selected predominately (sic) peatland types during those relatively mild winters. However, during the more difficult winter conditions of 1991/92, the caribou essentially abandon (sic) the peatlands and moved into stands on predominately (sic) mineral soil in the area north of the Chinchaga River. . . The area identified north of the Chinchaga includes some core caribou range for difficult winter conditions. The area south of the Chinchaga largely misses the caribou issue, containing some scattered sightings, predominately (sic) in summer range. While DMI maintains that protected areas are not the solution to caribou habitat management in themselves, they can provide a benchmark for caribou habitat, and should be considered during selection of a network of protected areas.

We recognize that there is significant energy sector development in the area north of the Chinchaga. It also needs to recognized that the in terms of meaningful time, (sic) the disturbance from this development is very temporary. We should have a longer term vision of the purpose of protected areas, and ensure that we are selecting a site which has the most value in the long term...

In regards to the process which has been used to narrow the field of candidate sites, it concerns me that this has been done independently by an inter-departmental government committee prior to any input from stakeholders. What particularly concerns me is the refinement of boundaries to a very small portion of the original nomination, with not even an indication of the original boundary, or the rational (sic) behind the elimination of portions of the nomination from consideration prior to input from stakeholders. It was my understanding that this would be the role of the Provincial Coordinating Committee, in consideration of input received... ”

THE SPECIAL PLACES
Provincial Coordinating Committee

One of the more controversial aspects of the debate over the Chinchaga centres on the discussion and recommendations of the Provincial Coordinating Committee (PCC) for Special Places. The committee is characterized by "balanced stakeholder representation" – its twenty-plus members represent a wide range of groups with an interest in establishing protected areas. As Ty Lund, the Minister of Environmental Protection said in 1995, the PCC is supposed to play a key role in recommending, designating, and establishing Special Places: "The role of the Committee is to provide consistent, overall direction for the program. This is critical to the success of the Special Places program. These representatives have a key role in helping increase public awareness and understanding of Special Places."[14] The following PCC opinions on Chinchaga are taken from the draft meeting minutes of the August and September meetings of the Committee. The excerpt refers to Level I themes. A Level I theme "is a broad landscape type within a subregion... Level I Themes can be used to portray the natural diversity within the various natural regions and subregions. Level I Themes are the most important classification level because of their close linking of landforms and dependent lifeforms."[15] Tom Mill, the Project Manager for Special Places in the Department of Environmental Protection, chaired the August 19, 1997 meeting.

I.U.

August 19/20, 1997 Provincial Coordinating Committee Meeting, Edmonton

August 19th: "T. Mill presented the information for the Chinchaga candidate site. A map was presented showing the proposed Grande Alberta Paper (GAP) Forestry Management Area (FMA). The map portrayed the effects on AAC with different candidate site options. An Energy map was also presented to show the Energy dispositions for the alternative Chinchaga boundary that was suggested at the July PCC meeting. CAPP (Editor's note:

Canadian Association of Petroleum Producers) did not object to either the original or the alternative site going to local committee because both represented equivalent Energy issues.

T. Mill suggested that the PCC go forward with the original candidate site because of the forestry conflict in the Halverson ridge area of the alternative boundary. The PCC members were undecided as to which boundary they wanted to put forward. It was suggested that a proposal would be drafted in the evening which would propose the alternative boundary that best meets the Level I themes with the least economic impact as well as recognizing the political sensitivities in the area. Discussion was deferred to the following day... "

August 20th: "T. Mill presented an overhead roughly outlining the PCC's suggested boundary configuration for the Chinchaga candidate site. The proposed candidate site captures part of the original candidate site and an area extending south towards and including part of Halverson Ridge. Suggested wording for the recommendation and considerations regarding the Chinchaga site were also presented which reflected lunchtime discussion by the Chinchaga subcommittee (Attachments 3 & 4). The PCC notes in their wording for the recommendation that their proposed candidate site better captures the Level I theme representation for the Foothills Natural Region. The PCC also recognizes the nature of the forestry commitments that led to the identification of the original candidate site. The AFPA (Editor's note: Alberta Forest Products Association) clearly indicated that forest tenure holders have not been consulted with respect to the recommended site. Other members (Coal Association, CAPP) had reservations about not having a refined boundary, however they agreed to support forwarding the Chinchaga site to a local committee.

*** Consensus to forward the Chinchaga candidate site recommendation for the reconfigured boundary to the Minister to establish a local committee. The candidate site boundary will be refined by the secretariat to represent the PCC's suggested boundary configuration... The words developed by**

the Chinchaga subcommittee will be incorporated into the recommendation. CPAWS registered their objection to the candidate site.

Attachment 3: Chinchaga Recommendation

The PCC recommends establishing a local committee to review the Chinchaga site.

PCC members recognize that government has important forest allocation considerations underway in the proposed G.A.P. Forest Management Area.

Based on our scientific, social and economic analysis of Chinchaga area lands, the PCC recommends that the Local Committee review a Chinchaga site that includes portions of the original candidate site and lands extending to the South, including Halverston (sic) Ridge.

We feel that this would provide a more representative site for local community consideration.

Attachment 4: Chinchaga Considerations

1. Change interim protection to apply to new site boundary and to be withdrawn from excluded parts of the present candidate site.
2. Local Committee representation and public involvement need to consider the wider social and economic spheres than are found within the M.D. (Editor's note: Municipal District) boundary."

September 23/24, 1997 Provincial Coordinating Committee Meeting, Drumheller

"Chinchaga

T. Mill advised the PCC that the Minister has accepted the PCC recommendation to move the revised Chinchaga candidate site to local committee review. This site boundary is as recommended by the PCC at the August PCC meeting."

C H A P T E R 8

PATHS TO CONSIDER

Canada's people are asleep for a nation,
That ought to be building its home on a rock.
Some day they'll awaken, or I'm very much
mistaken,
Some day, they'll get one hell of a shock.

Unwelcome citizens are grumbling,
Vandals are plundering.
And the elect of the people are shutting their
eyes.

- from "Canada's Weeping"
by Henry Stelfox[1]

— INTRODUCTION —

A few prescriptive thoughts conclude our look at environmental controversies in the Rockies and the Foothills. The controversies are obvious – are the solutions? The paths built in this final section are made from two ingredients: ideas and political leadership. Their signposts to a better future call for rethinking the relationship between the economy, technology and the environment and for a different style of political leadership.

Thomas Power and Mark Anielski offer the economic prescriptions. Power emphasizes the importance of environmental amenities, such as wilderness, to the locational preferences of people and business. Anielski urges governments to start to include the values of natural resources and the environment in their measures of economic growth and health. A conventional wisdom criticized by Anielski – technological change makes finite ecosystems infinite – is a focal point for Stan Rowe, the eminent ecologist. Rowe's examination of technology and ecology invites us to consider a vital question – what types of technology are compatible with a healthy ecology?

The Edmonton Journal and Brian Horejsi demand political leadership. Government must start to support the sincere efforts leaders in our business and environmental communities make to reconcile environmental protection with economic prosperity. Brian Horejsi's article, sparked by a controversial 1996 government decision to cull an elk herd southwest of Calgary by holding a mid-winter cow/calf elk hunt, asks the Alberta government to follow the example which a growing number of American states are following: Government purchases of ecologically significant lands. In June 1998, South Carolina, a state of more modest means than Alberta, finalized its purchase of

32,000 acres of wilderness in the Jocassee Gorges from Duke Energy. In Florida, public land purchases made through Preservation 2000, a $3 billion state bond program, enabled the state to save nearly one million acres from development. Will Alberta's politicians exhibit the political will needed to move in these directions or, like the preceding generation described by Stelfox, will they just shut their eyes? I.U.

SPECIAL PLACES A HODGE-PODGE[2]

- by The Edmonton Journal

It is not surprising that the Alberta government got a flunking grade from an international conservation group for failing to adequately set aside wilderness and natural areas.

The World Wildlife Fund's annual report card on provincial efforts to protect these areas has become a repeated embarrassment for the government, which received an F-mark three years ago as well.

At the top of the list of reasons for not making the grade has been the province's insistence that industrial activity should be allowed in land set aside under the Special Places program.

That's a program aimed at preserving examples of Alberta's wide variety of natural regions.

While 44 such sites have been established since the program's start three years ago, many contain activities such as logging, oil and gas development or mining.

While the government maintains that it must honour commitments it made to allow such development before the designation of such areas, conservationists rightly point out that preservation cannot co-exist with industrial activity.

In other provinces, allowing such activity in protected areas has become the exception rather than the rule, according to the Wildlife Fund.

In Alberta, the reverse is true. Clearcutting, oil and gas seismic lines, drilling and other intrusions into the landscape have become the rule rather than the exception.

Not that Albertans should get overly excited about the amount of land involved. While government officials crow about 1.2 million acres being set aside since the start of Special Places, that works out to 4,860 square km – 0.7 per cent of the province's entire area.

That meagre pace and the permitting of industrial uses are among reasons that one environmental group has pulled out of the Special Places program, while the remaining two are considering the same move.

Of course, some Albertans will dismiss this all as the ranting of a few special interest groups.

But some industries and progressive companies are also unhappy with the province's approach and are challenging the government over its vague yet inflexible approach to Special Places.

The Canadian Association of Petroleum Producers, for example, wants a better-defined program that sets out what industrial activity will or will not be allowed in these places.

It has even suggested a plan for quickly phasing out petroleum activity once an area has been set aside for protected status.

But the government has ignored that idea. It has left that question up to local committees, which often include other industry interests. The result has been a hodge-podge of so-called protected areas, some with more protection than others.

As for inflexibility, the province has been unwilling to free companies from industrial commitments – even when the companies themselves have requested alternatives such as land swaps.

A case in point. Husky Oil last year asked the province for a land swap after deciding that an area granted to it in Kananaskis Country was too sensitive for drilling.

The government refused, saying environmental restrictions on the activity would have protected the area, although part of it has been proposed as a Special Place.

So here we have a government ignoring a more sensitive hands-off approach to the environment from an oil company.

We have a government ignoring lobbying from an industry group that wants better rules over Special Places.

It's clear that cabinet ministers aren't listening to environmental groups on this issue. But if they won't even listen to the free enterprise sector that they profess to admire and promote, then who are they listening to?

It all underscores that big, fat F that has been awarded to the government.

HIGH QUALITY NATURAL ENVIRONMENTS
and Local Economic Development [3]

- by Thomas Michael Power

The Growing Importance of Residential Location Preferences

As the developing regional economies of North America move away from reliance upon primary production (agriculture, mining, and forestry) and manufacturing and become much more diversified, the population's preferences for living environments has come to play a more and more important role in the determination of where both population and business are located.

Consider the massive shift of population in the States from center cities to suburbs after the Second World War. This initially involved a move away from jobs and shopping opportunities and a significant increase in commuting costs. Why were these costs incurred? One major motivation was obviously to obtain a superior living environment: a less congested, more socially homogeneous neighborhood with "park-like" qualities (large lawns, single story dwellings, low population density, rural qualities, etc.). Or consider the more recent shift in population in the States from the "frost" or "rust" belt of northern industrial states to the "sun belt" of southern states. This was a move from high wage industrialized areas to low wage non-metropolitan areas. Here, too, qualities of the living environment played an important role in this migration of millions of people. In fact the term "amenities" was coined by a California economist in the early 1950s to explain the influx of population into the desert southwest despite the absence of any "economic base" to draw them there (Ullmann, 1955). Clearly climatic conditions was one element of the environment that played a role in the relocation of the population. But like the move to the suburbs, much of this move was an attempt to escape from the building legacy of environmental and social problems in the older, industrialized regions.

The impact of environmental amenities on the location of population and economic activity is not limited to these historical patterns. It continues. During the 1980s, most of the rural United States suffered a severe economic depression as agriculture, forestry, and mining all faltered. But a substantial set of nonmetropolitan counties escaped this serious downturn and actually prospered. The United States Department of Agriculture has labeled these counties "high amenity" counties because they are characterized by environmental qualities that have served as magnets that drew both population, employment, and economic activity despite economic decline elsewhere in the nation and rural areas. These attractive amenities included lakes, mountains, rivers, moderate climates, seashores, protected natural areas, cultural institutions (college towns), etc. (Deavers, 1989; Beale and Fuguitt, 1990)

Another example of this phenomenon is the impact of wilderness protection on adjacent communities. It is often claimed that protecting wildlands, because it limits extraction of natural resources, impoverishes communities by locking away valuable resources. But a survey of all of the counties adjacent to classified federal wilderness in the United States shows that for the last thirty years, these counties have had population growth rates that are two to three times the national average and the average for all nonmetropolitan counties (Rudzitis and Johansen, 1989 and 1991). Whatever wilderness protection has done, it has not choked off economic growth. One explanation for this growth which is supported by survey work is that people seek out high quality living environments and wilderness protection serves as an indicator of an area where a permanent commitment has been made to preserve the natural environment. This draws people and economic activity.

Note that the emphasis here is on the importance of high quality living environments to current or potential residents, not tourists. High quality living environments stimulate economic development simply by making a place an attractive place for people to live, work, and do business. Tourism, at best, is a sidelight to this phenomenon. One can expect this trend of people moving to where the qualities of the social and natural environment are attractive to continue. Environmental quality appears to be a "luxury" good. As incomes rise, more and more people choose to make sacrifices to obtains access to these qualities. Rising levels of education also seem to increase peoples' environmental sensibilities and lead them to consider environmental quality more closely when they make location decisions. Improvements in transportation and communication allow people to move some distance from large urban centers without facing economic or cultural isolation. Finally, as public and private retirement programs which were put in place 40 to 50 years ago begin supporting people who have contributed to those programs over their entire working lives, more and more of our population will enjoy "footloose" incomes which travel with them where ever they choose to live. This allows them to exercise locational choices free of economic constraints which, in turn, allows environmental preferences to play a larger role. All of these trends taken together suggest that environmental "amenities" will play an increasingly important role in the location of population and economic activity in the years to come. (Power, 1988)[4]

NATURAL CAPITALISM [5]

- by Mark Anielski

Since the advent of the Industrial Revolution our world has been transfixed on the mantra of economic growth – more production is assumed to be better. Economic growth has become an article of faith. Is it not time to debunk this myth?

In the fall of 1997 Alberta will engage in a public discussion, the Alberta Growth Summit, on how to continue developing the economy. While we talk about expanding growth, we do not know whether existing consumption is sustainable or if we might be compromising the quality of life in the process. Have we not simply been counting our pennies from the sale of our oil, gas and timber resources while paying no attention to the depreciation of our environment and natural resources?

No doubt we are measurably better off economically than our ancestors as we continue to expand the production of goods and services. Official economic statistics such as GDP (Gross Domestic Product) provide the evidence we need. Yet, most of us have a nagging feeling that our quality of life is eroding and our environment is being compromised. We have a sense, but little concrete evidence, that our environment is being polluted, our forests depleted and our oil extracted at unsustainable rates. We do not know if such a path is sustainable for ourselves or future generations. A useful analogy is that of an automobile called "economy" hurtling towards a cliff with the chauffeur (the stock market, politicians, the media) transfixed on a select set of dashboard indicators. The instruments are either broken or flawed in their design. We can't tell if we are out of oil, have engine trouble, or if potholes, or even blackholes, loom ahead. Instead, we drive along a growth curve in a blissful state, oblivious to the potential defaults that lie ahead.

How did this economic growth journey begin? GDP and the System of National Accounts (SNA) were first conceived by a few United Nations economists following World War II. At that time the world was focused on rebuilding shattered economies, and GDP and income accounts were established to measure economic progress. GDP simply measures the value of a nation's annual production of goods and services. Every time a good or service is bought or sold the transaction contributes to the annual GDP figure. GDP is the most important measure of an economy's health. The smallest change can have profound effects on financial markets and the economic policies of nations.

What GDP and the SNA do not measure is the value of the environment, natural resources and so-called "women's work." This may not be new to most economists, but the average person is not award of such truths. Marilyn Waring, economist and former New Zealand MP, has made the exposure of this truth a personal crusade. Waring found to her dismay that the handbook for economic accounting (the SNA) in the UN headquarters in New York explicitly states that the environment, natural resources and women's work are not to be included in an economic account. Waring, along with other economists, including Herman Daly, Robert Repetto and Clifford Cobb, has attempted to draw attention to the shortcomings of

GDP and to adjust traditional measures to take into account the depreciation of nature's capital.

Bottom line: "nature's capital" including both renewable (forests, fisheries, farmland, wildlife, wind) and non-renewable (oil, gas, coal) resources plus environmental services that provide clean air and clean water are not accounted for in measuring the overall health or growth of our economies.

But so what? What are the consequences of excluding the environment from our economic accounts? Take the Exxon Valdez oil spill, for example. This disaster resulted in tremendous economic activity including the clean up of the spill, the legal disputes and the economic studies to assess the damage. All these transactions counted as a measurable increase in the U.S. Gross Domestic Product. Ironically, GDP rose more because of the spill than if the Valdez had safely delivered its load. Yet, nowhere in the U.S. economic or public accounts is there an account of the economic, physical or qualitative loss to the Alaska fishery, wildlife, marine ecology, tourism or to local communities. An increase in the GDP was the only evidence in the U.S. of the Valdez disaster.

The same is true of Alberta's economic accounts. Provincial economic accounts and GDP provide no account of the inventory, the value or the depreciation of our "natural capital." Yet these natural capital assets contributed roughly 40 percent to Alberta's GDP in 1995. Ignoring the value of these natural assets is akin to a steel company ignoring its inventory of steel and not accounting for the depreciation of physical plant and equipment. So why is nature's capital treated differently?

Paul Hawken's recent article titled "Natural Capitalism" (*Mother Jones*, March 1997) explains that part of the reason lies in the assertion of most economists that "natural and manufactured capital are substitutable – that we can invent technological adaptations to compensate for the loss of living systems." We assume that every time we run out of nature's capital stock we can create a perfect human-made substitute or find an alternative natural capital substitute. This economic law suggests that the economy can expand exponentially. Yet this seems counter-intuitive. As Dr. Herman Daly, former senior economist at the World Bank, asks what good is a sawmill when you have run out of timber, or a fishing boat when the fish stocks are depleted?

When Daly was at the World Bank, he attempted to draw a picture of the "full-world economy" as a box expanded to the outer limits of a circle called the "ecosystem." His effort was rejected. Such a picture would have created a crisis in economic circles since it would have been an admission that the economy does face limits to expansion, bounded by the capacity of the ecosystem. The ecosystem is in turn dictated by the laws of thermodynamics. Physics teaches that the total supply of energy in the system called earth cannot change, thus our physical environment is in a steady state. All that humans can do is manage their own consumption of nature's capital in a way that does not preclude consumption by either current or future generations. The failure of the Biosphere Project II in Arizona teaches us that we cannot manufacture a self-sustaining ecosystem as a substitute for the natural environment, regardless of our ingenuity and money.

A few economists, including Daly, Repetto and Cobb, have designed alternative economic measures that can assess sustainable development. Such measures would take into account the environment and natural capital. Repetto and other authors, including myself, have developed natural capital accounts that account for the changes in the physical, qualitative and economic value of nature's capital. Cobb and others have designed a Genuine Progress Indicator (GPI) – a revised form of GDP – that incorporates the negative costs of environmental degradation, crime, loss of leisure time and the loss of other quality of life issues that can be quantified.

The natural capital accounts I have helped develop for Alberta attempt to reveal evidence as to the state of our natural capital. This requires a physical inventory of both quantity of resources and quality of environmental services, as well as the valuation these assets (if value can be discerned).

Preliminary accounts for Alberta's timber reveals that the timber capital in our forests has remained sustainable from 1962 to 1995. A timber sustainability index (TSI), calculated as the ratio of total estimated growth of Alberta's timber capital to total depletions, reveals that sustainability has never been compromised... In every year since 1962 the volume of timber harvested has never exceeded annual growth of the forest. Since 1962 only about 18 percent of Alberta's 22.5 million hectares of productive forest land has been harvested, burned or infested.

Accounting for fixation of carbon shows that Alberta's forests and peatlands absorb roughly 24 percent of the 1995 emissions of carbon dioxide (CO2) from Alberta's petroleum industry. This does not account for the additional emissions from automobiles and other sources. Alberta is effectively a net exporter of carbon. Also, because Alberta's forests are essentially in a steady state over time, in accordance with the laws of thermodynamics, the capacity to absorb carbon is also constant. Thus, while emissions of carbon from fossil fuel production continue to rise, there is little we can do to counterbalance these rising emissions by "managing" our forest environment.

Alberta's oil, gas and coal accounts reveal a steady decline in reserve life of conventional crude oil and natural gas... The reserve life for conventional crude oil was 6.9 years in 1995 and 12.1 years for natural gas (reserve life is an estimate of the years of economic life remaining for a non-renewable resource). Alberta's oilsands are substantial with an estimated ultimate reserve life of over 3,000 years while coal reserves would last well over 1,000 years at current production.

To date, no accounts exist for agricultural soil quality, wildlife, biodiversity, air quality, and water quality and quantity.

To make the appropriate adjustment to GDP, we need to convert the physical natural capital account into a monetary equivalent. This is done by estimating the annual economic rent (the value of a unit of natural capital to the province) of natural capital and multiplying it by the annual physical account figures. Only the forms of nature's capital that are traded in a market can be valued in traditional accounting terms. Many environmental services and resources have no market value – air, water, wildlife, biodiversity, carbon fixation, air emissions. With the monetary account information we can estimate the value of a change in the quantity or quality of natural capital and thus adjust our GDP figures to reflect their depreciation. We end up with a "green-GDP" figure, a better measure of sustainable development of Alberta's natural resources.

The monetary account reveals the value of natural capital and provide information to assess the returns to natural capital consumed for economic development. The timber account reveals that a tree may be at least as valuable left standing and absorbing carbon dioxide than if it were harvested for lumber or pulp. The account also reveals that the economic rents generated per cubic metre of timber harvested may, in certain cases such as lumber, be negligible and, in some cases, negative. If one considers stumpage/royalty fees and corporate taxes paid by the forest industry as a return to the province for each cubic metre of timber, the returns may be questionable especially when compared to the public costs of fire fighting, forest management and infrastructure that primarily benefit the industry. Add to that the costs of failed financial assistance, such as the Millar Western loan, and the returns look even less favourable. Such evidence raises serious questions as to the most efficient and effective use of publicly owned natural capital. Is a hectare of forest land best used for timber harvesting, for carbon sequestration, oil and gas production, agriculture or as a park?

The monetary account for oil and gas allows us to determine whether sufficient economic rent was captured through royalties. The account can also serve as the basis for assessing how much of the royalty revenues should have been reinvested, rather than spent, in order to sustain Alberta's overall wealth. The goal with non-renewable natural capital is to invest a portion of the rents in a renewable form of energy capital (e.g. hydro, solar or wind energy) or financial capital (e.g. Heritage Savings and Trust Fund) to ensure a sustainable income stream over time that offsets the loss of royalties when the non-renewable energy is depleted.

Paul Hawken suggests that natural capitalism requires:

- An ethic towards the environment that accounts for the state and value of nature's capital.
- An attitude that it is cheaper to take care of something by maintaining and repairing it than to let it decay and try to fix it later, especially when that "something" is as irreplaceable as some environmental services.
- A revision of our tax laws that recognizes the value and role of nature's capital.

What we need to do now is walk the attitude talk.

TECHNOLOGY AND ECOLOGY[6]

- by Stan Rowe

Technology has become the primary means by which humans interact with the Earth-home. An outcome of artifice guided by human beliefs and purposes, technology makes visible the deep beliefs that consciously and subconsciously motivate society. Through it the people/planet relationship is made explicit. Hence the importance of a critical appraisal of technology, of its effects on the world and of the extent to which its various forms are appropriate.

Technology, the Big T, has been variously defined. All agree that it is more than hardware, more than machines, tools, material instruments. An inclusive definition of T is: a *reproducible and publicly communicable way of doing things*. The key word *communicable* shifts T into the world of ideas, language, beliefs, culture...

An essential part of T consists, therefore, of cultural ideas, especially the values and goals espoused by influential people – in our society by business people. Their materialist philosophy provides the inspiration and encouragement for industry's material production. The built environment, steadily expanding, is the visible expression of T's cultural authority, influence and growth. Progress is T, and T is progress.

Technology as Progress

Faith in T is deep rooted. Aldous Huxley considered it the modern religion, its symbol a decapitated cross and the Model T Ford. T is our provider and in it we trust to give us our daily bread and deliver us from evil. Yet the Good Life so provided is not free. Environment has been picking up the tab, paying our way. Now its ability to support humanity's appetites and wastes is running out.

The conventional use of the word "progress," often preceded by a resigned or cynical "you can't stop it," brings to mind T's gifts to our consumer society: living better electrically, acquiring more machines, eating exotic foods, travelling faster and perhaps some day visiting the moon and planets. Progress is living "high on the hog" for the largest possible percentage of the human population, using more and more energy and processing more and more materials from the earth's crust. Will we continue to call it progress as it slowly kills us?

Progress in a finite ecosystem cannot mean the absolute ascendancy of one species. Progress must mean the achieving of a creative symbiosis within the Home Place, where sympathy and care extended by the dominant species to the rest of creation. To transcend traditional preoccupations with our own kind, preparing to appreciate the Ecosphere with all that is in it, requires an understanding of T and how we have used it both to exploit and to distance ourselves from Mother Earth.

Exosomatic Technology – Appendages for People

Technology churns out a variety of artifacts: tools such as pile-drivers, media such as television, cocoons such as buildings and cars, all of which like spectacles and hearing aids can be hooked on the body to increase the wearer's abilities and pleasures. Aircraft are fast legs, attached by seat belts to people in order to increase their speed of travel. Submarines are worn to swim under the oceans, microscopes and telescopes to see better and farther, telecommunication systems (giant vocal chords) for long-distance information exchange, bulldozer and dragline are muscles for earth moving, computers extend the mind's scope. Transmitted from generation to generation by culture rather than by genes, T is the new means of human evolution, increasingly adding to human versatility, power and size.

From this perspective, T provides (in addition to biomedical inside-the-body gadgets) a wide range of outside-the-body or exosomatic instruments, extending human abilities to change the world and all that is in it. Side effects are extraction of vast quantities of "raw materials" from the planet's surface to manufacture the array of technologic products, and a return flow of wastes and poisons at all stages from primary production to obsolescence.

Initially purpose shapes the instruments, but soon, like body parts, they begin to inspire their own use. Just as those born with good vocal chords can with difficulty be dissuaded from singing, so a boy with a new hatchet finds much that needs chopping. The United Nations dictum, "Wars begin in the minds of men" is a half-truth. Wars are also encouraged by the availability of arsenals of shiny potent weapons whose purpose is destruction. Dis-armament is effective because it amputates the striking arms, separates the muscle-toys from the aggressive boys, removes the temptation to let the exosomatic instruments "do their thing" at the slightest provocation.

If technological instruments are body extensions, may they not also be viewed as disfiguring outgrowths, as excrescences with pathological tendencies? Three potential afflictions we suffer from "wearing" our technological armour are gigantism, addiction and alienation.

Technology and Gigantism

All technologies whose goals are increased power and control effectively lead to gigantism. Machines behave like enlarged limbs and organs, demanding energy for growth, repair and reproduction just as organisms do. A North American, considered together with her exosomatic appendages, is 80 times the size of a Bangladesh peasant judged by appetite for non-renewable resources (fossil energy, minerals) and renewable (ecosystem) resources.

Each Western person bestrides the world like a Colossus, leaving his Sasquatch imprints on land and water through the massive extraction, consumption and waste of resources that enlarged body size necessitates. Head counts are misleading when technologic size is not factored in. Is Canada underpopulated with 26 million people? Before answering, bring in the technologic multiplier, the energy use per capita, that indicates our impacts on environment. Considering that every Canadian stomps the Earth like 80 peasants, our country is vastly overpopulated. Further, in a world of rapid population growth but fixed space, of polluted renewable resources and dwindling non-renewable fossil fuels, the expectation that everyone can be an exosomatic giant on the North American model is unrealistic. The dream of the world's poor, who make up the majority of the human race, casts over the future a dark cloud that will only be dispersed when the rich nations begin to share by divvying up their wealth. Sharing is more realistic than depending on economic growth, for the world will never support, at Canadian standards, the six billion people expected in the year 2,000. One billion could perhaps live as well as the citizens of Calgary and Winnipeg. Some say only 250 million.

The gigantism that industrial T confers is largely material. It effectively extends the power of our material bodies without doing much for our immaterial minds. Like a growing brontosaur, T's body size has far outdistanced brain size. Encouraged to be consumers, we surround ourselves with more and more artifacts, increasing our appendages and faculties, while our reasoning power and modicum of wisdom stay the same or shrink before the TV set. The distortion of the body/brain ratio creates feelings of helplessness. The common fear that machines are out of control, that they are running people's lives, suggests that exosomatic evolution has outstripped its rational management.

If the diagnosis is correct, we would do ourselves a favour by reducing exosomatic T to restore a better body/brain ratio. In industry, fewer mega-projects and more "soft" local developments would reduce our gigantism. In agriculture, a larger eye-to-acres ratio would improve on the wheat-belt trend of fewer eyes on larger farms.[7] Instead, the favoured prescription is to keep T growing, not attempting to enlarge comprehension per se but to supplement and beef up brain power with the latest T: AI, Artificial Intelligence, that also doubles for Artificial Insemination. If only T could deliver AW, Artificial Wisdom!

Technology as Addiction

Because they give us power and pleasure, exosomatic instruments are addictive. When "worn" even for a short time they become necessities, difficult to lay aside unless replaced with better, more efficient models. The rancher turns in his horse for a half-ton truck. The youth moves up from motorcycle to musclecar. For those habituated to fast movement through the air, surface travel – except perhaps for recreation – loses its charm.

Progress is traditionally conceived of as greater power, control and efficiency – not less. To suggest giving up powerful Ts on the suspicion that they are dangerous and destructive is to invite scorn and accusation of wanting "to go back to the cave." Georgescu-Roegen thought that this addiction – as attractive to humanity as a flame to moths – is the greatest obstacle to a rational human ecology. He was pessimistic about cures for the pervasive dependence on exosomatic comforts. "Perhaps the destiny of man is to have a short, but fiery, exciting and extravagant life."[8]

Like addictions to drugs, sex and gambling, being hooked on T is controllable if not curable. The first step is recognition that we have a problem, that T is a predicament, which then suggests the need to understand it and find satisfying alternatives to its destructive facets. The sometimes brilliant philosopher William James thought that the moral equivalent to warfare was an all-out battle with Nature, a constructive alternative to killing each other! We have proved that such a battle is an immoral equivalent, for it kills everything. Perhaps a moral equivalent to war can be found by distinguishing friend from enemy, separating appropriate T that assists our symbiosis with the universe from the poisonous and destructive T of heavy industry and military might, then launching an all-out attack on deadly kind before it destroys the world and ourselves completely. Let us declare our debilitating addiction to instruments of power and explore the possibilities of cutting Big T down to little t.

Technology as Barrier

T is also an alienating influence, interposed between people and the planet. As the manufactured milieu grows, contact between people and the Ecosphere is steadily reduced. The built environment acts as a filter and barrier that "progress" renders thicker and more opaque, gradually eliminating and shutting off nature's direct sensual stimuli. During the last eclipse of the sun over southern Saskatchewan, children in many schools were only allowed to view that celestial phenomenon on television. Watching TV in buildings without windows is the model of self-induced alienation, annihilating all sense of ecological roots and dependencies.

T has been described as arranging the world so as to minimize direct experience of it, an alienating effect that results in sensory deprivation. Cut off from Nature's sights and sounds, people end up in single-species confinement. Aloneness in the city, with nothing to sense but themselves and made-things, induces various psychoses. T's cages isolate people from the Ecosphere milieu – their biological and evolutionary home – as effectively as iron cages isolate lions from their savannas. Just as zoo animals deteriorate when deprived of their natural home surroundings, so also humans as earthlings deteriorate when, like the legendary Antaeus, they lose touch with planet earth.

Technology as Placenta

From the standpoint of human ecology, T is the system of values, beliefs and techniques by which a society, a culture, taps into, uses and modifies the Ecosphere, in the process changing both itself and the surrounding system. T is the instrument of contact between people and the miraculous enveloping environment from which over four and one-half billion years of evolution they came. T is the means by which humanity extracts what it wants from the earth and returns what it does not want to the earth, playing an analogous role to the placenta and umbilical cord connecting fetus and mother. Should the fetus imperil its host by excessive demands and generation of wastes (a likely effect of gigantism), then both will suffer. As with mother and fetus, the only safe and sustainable ecological relationship between people and planet is one of moderate demands in a symbiotic alliance.

The dependency of people on the Ecosphere is complete. Indeed, all organisms are inseparable from their environments except in thought, and life is more an attribute of the Ecosphere than of the organisms it encapsulates. The natural world and all forms of life within it are interpenetrated. Sophisticated T does nothing to reduce the people-planet dependency, but it does magnify the possibilities for exploiting and poisoning the relationship. The onus is on humanity to develop a technology appropriate to the well-being of the earth, a T that does not imperil the Ecosphere but contributes to its healthy functioning.

Conventional T is purposive and aggressive. Comprising values and ideas as well as tools, it urges its own uses which, given our cultural past, are people-serving in a short-sighted way. Unless we recognize this fact, we slide into the popular "technology is neutral" mode of thinking that backs away from any kind of control over our inventions and shrugs off the need for ethical choices at T's leading edge. The ultimate nonsense is justification of war because of the technologic advances it brings – poison gases that later yield insecticides, tanks that are the models for better tractors. Danger lurks in the idea that nothing is wrong with any T, that it is only what we do with it, what we make of it, that can be judged ethically. This, said McLuhan, is the numb stance of the technological idiot.

Evaluating Technology

Many human problems are unsolvable in the humanistic context. What is more important, mother or fetus? If human life, actual and potential, is believed to be paramount, no conclusive answer is possible. But in the context of the Ecosphere, knowing what we know about carrying capacity and of social and environmental destruction due to overpopulation, most ecologists will come down on the side of the mother.

Similarly with T. What is appropriate and what inappropriate cannot easily be decided if the arguments consider only help or hindrance to people. Will it feed more of them, will it prolong their lives, will it make them more powerful, will it help them calculate faster? The usual conclusion is yes, do it, because people will be "better off."

A sounder standard for judgement is ecological, measured against the requirements for symbiotic survival in the Ecosphere. Then the measure of T's goodness or badness is its effects on the Earth system from which it draws materials to which it returns wastes. Hitherto progress has meant more T, more industrial growth, more material goods, more wealth at the expense of the Ecosphere. Progress must be redefined as sustainable ecological relationships between humans and the Ecosphere, to which new kinds of T evaluated in the context of health and permanency can contribute. That is, T for two.

Appropriate T will respect the vital milieu of air-water-soil-sediment-organisms. It will not appropriate these earth-surface resources faster than they are replenished and renewed. Nor will it introduce into the life-space pollutants and toxins from underground. Of the latter, in particular, it will not play around with radioactive substances nor falsely promote them as safe, clean and cheap. Rather than encouraging nuclear indigestion, T will help the world avoid atomic ache.

THE NEED FOR A PUBLIC LANDS STRATEGY[9]

- by Brian L. Horejsi

I recently had a relatively rare conversation. Rare, at least for me, because it was with a provincial politician. He asked me, What about this Horejsi? What about this elk "hunt" south of Calgary, and what about the wolf slaughter in southern Alberta over the past year? So I told him "what" about it! Here's what I had to say.

The two are peas in the same pod, that pod being land use issues. No surprise there. This is a battle over resources and who controls them. But more than that, it's about the diversity of wildlife populations and the ecological integrity of public lands and ecosystems that overlap private land. It's about public ownership and public control and public responsibility for wildlife, land, and tax dollars. And it's about solutions to these chronic flare-ups that have plagued Albertans from before provincehood. Simply put, this reaches all the way to the roots of democracy and whether the majority of Albertans can influence the survival of ecosystems we all need for our survival.

Albertans have said time and time again that they want wildlife protected; they say it in polls, in letters to the editor, by going to court, and by gobbling up every ticket to hear David Suzuki. Hunters also want wildlife. some of them see that the viability of a wildlife population is a matter of ecological integrity and human integrity, often theirs to be exact; and I give praise to those who have stepped to the fore in recent weeks and spoken about what hunting is. It is not about killing animals whose fetuses are now over a foot long (30 cm) and weigh over three pounds (1400 grams). And hunting is not about killing animals that are concentrated and weakened by severe weather and who have adapted to, and would normally be in, a reduced state of physiological and behavioral activity to, ironically, improve chances of overwinter survival. Honest wildlife management is not issuing licences to kill highly vulnerable elk from a population of about 15,000 animals province-wide when the provincial management goal is 30,000 animals.

Hunting is ethical, law-abiding pursuit under fair chase rules. But hunting is also only part of ecologically sound management – the biggest part of which is maintaining the biological integrity of the wildlife community: natural sex ratios (not eight males per 100 females when they're born close to 50:50), genetic variability, natural predator prey relations, and the full complement of natural processes such as migration. These things require minimum population sizes, and to do that we require a land base. Land on the prairies, land in the foothills, and land in the mountains! Land with a high degree of habitat effectiveness, a quality that allows annual physiological and behavioral requirements to be met. And latitude – space – to accommodate tough winters or perhaps a major fire (coming soon to most of Alberta's forests), or even (for bears) berry crop failures.

In both the elk kill area and southern Alberta there will be no solutions to the conflicts, just treatments of management failure. But even more unfortunately, there will never be a complete native wildlife community to provide the economic and social benefits that Albertans could use. At least there will be no solutions with (1) the present public land base, (2) the fragmented state of public lands in the foothills, and (3) the present state of degraded habitat effectiveness on all public lands.

The solution to these recurring confrontations, while not simple, is obvious: acquire land for public ownership and consolidate public land into ecologically functional units. This will require some claw back, a term taxpayers are familiar with – taking back some of the space, resources, and subsidies that public land users now enjoy at taxpayer expense. And it will require the willing-seller purchase of ecologically significant and substantial blocks of land starting, I would suggest, with some critical areas along Alberta's foothills. These will have to be wildlife ranges that are large enough to provide essential ecological options like relief in land form, thermal and security cover, and space to accommodate movements and limit peripheral conflict with private land owners. These lands, in conjunction with existing public lands (i.e., the Green zone), must have the capability to maintain viable populations of ungulates (deer, elk, moose) for utilization by predators, including (sometimes) hunters, and populations that have the capacity to absorb management mortality.

In the real world, land acquisition for wildlife is a common event. When the state of Montana bought land and established the Sun River Game Range in the early- to mid- forties, only 36 percent of the elk wintered in the area but, with inducements, they learned quickly. By 1949-50, 79 percent of the elk in the population, now much larger than at the time land began to be assembled, wintered on the state-owned land. This area (20,000-plus acres) has become an ecological gold mine in the almost 50 years since then, acting as an ecological umbrella for all species.

Conservation biologists have known roughly what's required in southern Alberta, but what about the $50 million that we need to deposit in a preferably independent habitat Land Heritage Secretariat? We're fortunate to be in a province with multiple solutions to this problem. There are lottery funds – over $500 million per year available; how about one-tenth for one year? The Klein government claims income has been exceptional this year; imagine returning a relatively small portion to the land and wildlife most Albertans feel so strongly about. Or perhaps a penny per unit heritage levy on natural gas or crude oil exports? After all, each road, each wellsite, each right-of-way detracts from wildlife habitat. Giving a little back to our natural heritage seems long overdue.

Perhaps the best source would be redirecting surface access payments, now going into the pockets of private "holders" (lessees) of public lands, into the Land Heritage pot! Although the vast majority of the money would be spent on land, not all of it would be. An aggressive crop and haystack damage prevention program on adjacent lands; a manager, assembling a database, keeping track of events and working with local people, would be part of the package.

Even though time is running out, it would be a treat, wouldn't it, to join the 1990s by solving a century-old problem with a solution that would go a long way toward restoring a sense of fair play and serving the public interest ... doing the right thing that is ecologically sound and necessary and would, both in the short term and in the final analysis, be politically honorable.

NOTES TO THE CHAPTERS

PREFACE

1 William Butler Yeats,"The Lake Isle of Innisfree", Poems. Richard Finneran, ed. (New York: Macmillan, 1983).

CHAPTER 1

1 This poem appeared in Borealis, Vol. 2, no. 2 (Fall 1990). It is reprinted with the permission of Mrs. Barbara Whyte and The Canadian Parks and Wilderness Society.

2 Alberta Environmental Protection, Recreation and Protected Areas Division, Selecting Protected Areas: The Foothills Natural Region of Alberta, Appendices, (July 1996), p. 17.

3 Alexander Henry, New Light on the Early History of the Greater Northwest. The Manuscript Journals of Alexander Henry and David Thompson, Volume II, (Minneapolis: Ross and Haines, 1965), pp. 677-678.

CHAPTER 2

1 This poem originally appeared in Borealis, 2: 2 (Fall 1990). It is reprinted with the permission of Mrs. Barbara Whyte and the Canadian Parks and Wilderness Society.

2 As quoted in Andrew Nikiforuk, "The Nasty Game:" The Failure of Environmental Assessment in Canada, (Toronto: Walter & Duncan Gordon Foundation, January1997).

3 This article originally was published in Alberta Report, 12 January 1998. It is reprinted with the permission of Alberta Report.

4 Excerpted with permission from Leaning on the Wind: Under the Spell of the Great Chinook by Sid Marty copyright © 1995 by Sid Marty. Published in Canada by HarperCollinsPublishers Ltd.

5 Editor's note: the ERCB has been reconstituted as the Alberta Energy and Utilities Board.

CHAPTER 3

1 This poem originally appeared in Jeanne Perreault and Sylvia Vance (ed.), Writing the Circle: Native Women of Western Canada, (Edmonton: NeWest Press, 1990). It is reprinted with the poet's permission.

2 "Oil companies spend record amount for exploration," The Edmonton Journal, 10 January 1997.

3 Andrew Nikiforuk, "Hare-Brained Schemes," Canadian Business, February 1995.

4 Ibid.

5 "Industry, Green Groups in Canada Harden Positions," Platt's Oilgram News, 72:188, 28 September 1994.

6 This article was published in Environment Views: Alberta's Magazine on the Environment, 17: 2 (Winter 1994). It is reprinted with the editor's permission.

7 Editor's note: Alberta's known remaining gas reserves are shrinking. At the end of 1997, Alberta's remaining natural gas reserves stood at 1284 billion cubic metres,

making Amoco's Whaleback estimate equivalent to more than three per cent of known 1997 reserves.

[8] Alberta Environmental Protection, Alberta's Montane Subregion, Special Places 2000, and the Significance of the Whaleback Montane, (November 1995), 36-39.

[9] Alberta, "News Release - Local Committees Asked To Review Special Places Sites In Grasslands and Foothills Natural Regions," September 19, 1997. My emphasis.

[10] James Tweedie, "The Whaleback - Why We Left the Talks," GreenNotes: The Newsletter of the Calgary/Banff Chapter of the Canadian Parks and Wilderness Society, 7: 2 (Spring/Summer 1998). Reprinted with the permission of the author.

[11] Alberta Special Places Whaleback Local Committee, Recommendations to the Minister of Environmental Protection (Draft Report), (April 28, 1998), 13, 15.

[12] "Oil firms quietly conserving land," The Globe and Mail, 2 June 1998.

[13] The complete title of this document is "Quirk Creek Oil and Gas Lease and Expanding the Elbow-Sheep Wildland Park." This unsigned commentary was prepared by the Alberta Wilderness Association in November 1997.

CHAPTER 4

[1] This excerpt is reprinted with the permission of the Stelfox family.

[2] Alberta Energy and Utilities Board, Transcripts of EUB-CEAA Joint Review Panel Hearing Examining the Cheviot Coal Project, (1997), 800.

[3] This article originally appeared in Borealis 1: 4 (Winter 1990). It is reprinted with the permission of the author and the Canadian Parks and Wilderness Society.

[4] Editor's note: In 1998, the provincial government will pay 85 per cent of the commercial market value of cattle which are confirmed grizzly kills. Fifty per cent is still paid for a probably grizzly kill.

[5] This article originally was published in Encompass: Alberta's Magazine on the Environment, 2: 1 (1997). It is reprinted with the permission of the author and Encompass.

CHAPTER 5

[1] This poem is reprinted with the poet's permission from Monty Reid, The Alternate Guide, (Red Deer: Red Deer College Press, 1985).

[2] Editorial, The Hinton Parklander, 6 January1997. Reprinted with the permission of The Hinton Parklander.

[3] Alberta Energy and Utilities Board, Transcripts of EUB-CEAA Joint Review Panel Hearing Examining the Cheviot Coal Project, (1997), 2771.

[4] These excerpts are reprinted with the permission of Mayor Ross Risvold.

[5] Editorial, The Hinton Parklander, 24 February 1997. Reprinted with the permission of the Hinton Parklander.

[6] Alberta Energy and Utilities Board, Transcripts of EUB-CEAA Joint Review Panel Hearing Examining the Cheviot Coal Project, (1997), 449.

[7] Alberta Energy and Utilities Board, Transcripts of EUB-CEAA Joint Review Panel Hearing Examining the Cheviot Coal Project, (1997), 2758-2760.

[8] The Edmonton Journal, 7 June 1997. Reprinted with the permission of Linda Goyette and The Edmonton Journal.

[9] This article originally appeared in Encompass: Alberta's Magazine on the Environment, Vol. 1, number 2 (May 1997). It is reprinted with the permission of the author and Encompass. Dianne Pachal is the conservation manager for the Alberta Wilderness Association.

[10] At the end of June 1998 the World Heritage Committee considered a report on the Cheviot mine outlining the mine's approval process and Ottawa's assurances that steps would be taken to mitigate the environmental damage the mine will cause. The Committee accepted these assurances but asked to be briefed on the status of the mine in the late summer.

[11] The signatories were: David Schindler, D. Phil,. D.Sc., FRSC, Killam Professor of Ecology; Suzanne Bayley, Ph. D., Associate Prof., Biological Sciences; Dale Vitt, Ph. D., Professor, Dept. of Biological Sciences; John Packer, Professor Emeritus, Dept of Biological Sciences; Vince St. Louis, Ph. D., Assistant Prof., Biological Sciences; Jackie Huvane, Ph. D., Research Associate, Biological Sciences; Jules Blais, Ph. D., Research Associate, Biological Sciences; Jim Butler, Department of Forest Sciences; John Terborgh, James B. Duke Professor, Duke University; Frank Wilhelm, M. Sc., Grd. St., Biological Sciences; Michelle Bowman, M. Sc., Grd. St., Biological Sciences; Maggie Xenopoulos, M. Sc., Grd. St., Biological Sciences; Bill Donahue, B. Sc., Grd. St., Biological Sciences; John Clare, B. Sc., Grd. St., Biological Sciences; David Kelly, B. Sc., Grd. St., Biological Sciences.

[12] Alberta Energy and Utilities Board and Canadian Environmental Assessment Agency, Report of the EUB-CEAA Joint Review Panel, Cheviot Coal Project Mountain Park Area, Alberta, (June 1997), 75-76.

[13] Ibid., 79-80.

[14] Ibid., 81.

[15] Ibid., 88-89.

[16] Ibid., 131.

[17] Ibid., 131-132.

[18] Ibid., 132.

[19] Ibid., 132-133.

[20] Ibid., 143.

[21] Ibid., 143-144.

[22] Ibid., 144-145.

[23] Ibid., 160-161.

[24] Editorial, The Edmonton Journal, 23 June 1997. Reprinted with the permission of The Edmonton Journal.

[25] Editorial, The Hinton Parklander, 7 July 1997. Reprinted with the permission of The Hinton Parklander.

CHAPTER 6

1 The Banff Crag and Canyon, 31 August 1988. Reprinted with the permission of The Banff Crag and Canyon.
2 This article appeared originally in Environment Views, 17: 2 (Winter 1994). It is reprinted with the permission of the author and the editor of Environment Views.
3 Banff-Bow Valley Task Force, Banff-Bow Valley: At the Crossroads, (Ottawa: Minister of Supply and Services Canada, 1996), 150.
4 Ibid., 217.
5 Ibid., 252.
6 "Town says biz as usual," The Banff Crag and Canyon, 24 September 1997.
7 "Copps says Banff town plan is a no go," The Banff Crag and Canyon, 10 June 1998.
8 Canada, House of Commons, Debates, June 10, 1998, 1455.
9 "Banff mayor critisizes (sic) Copps' tactics on town plan," The Edmonton Journal, 22 June 1998.
10 "Ottawa unveils Banff growth plan," The Globe and Mail, 27 June 1998.
11 The Globe and Mail, 20 September 1997. Reprinted with permission from The Globe and Mail.
12 The Edmonton Journal, 11 June 1998. Reprinted with permission from The Edmonton Journal.
13 "Locals 'saved' Banff for future," The Edmonton Journal, 27 May 1998.
14 "Banff project bad for bears," The Toronto Star, 29 June 1998.
15 "Feds give nod to Chateau expansion," The Banff Crag and Canyon, 20 May 1998.

CHAPTER 7

1 This excerpt is reprinted with the permission of the Stelfox family.
2 "New Government Document Provides Framework for Sustainable Forest Management in Alberta," Government of Alberta Press Release, 26 February, 1998.
3 The Act is a reference to the Canadian Environmental Assessment Act.
4 Andrew Nikiforuk, "The Nasty Game:" The Failure of Environmental Assessment in Canada, (Toronto: Walter and Duncan Gordon Foundation, January 1997), 8.
5 "Court ruling 'ludicrous' says Lund," The Edmonton Journal, 13 July 1998.
6 Alberta, Environmental Protection, The Alberta Forest Legacy, no date, 5.
7 Sunpine's words here are identical to those written by the Alberta Conservation Strategy's Steering Committee in their October 21, 1995 draft of The Alberta Conservation Strategy. See page twelve.
8 Sunpine Forest Products, "An Overview of the Detailed Forest Management Plan proposed by Sunpine Forest Products," November 19, 1996, 18-20.
9 Hamish Kimmins, Balancing Act: Environmental Issues in Forestry, (Vancouver: University of British Columbia Press, 1992), 72.
10 The article from which these excerpts were taken appeared in Alberta Forest Products Association, Forestline, (March 1991).

[11] This excerpt is reprinted with the permission of Dr. Kimmins. An excellent overview of environmental issues in forestry is found in his book Balancing Act: Environmental Issues in Forestry, 2d edition, (Vancouver: University of British Columbia Press, 1997).

[12] Alberta, Department of Energy, "Information Letter 97-1," 6 January 1997.

[13] Canadian Parks and Wilderness Society, "Press Release," 22 April 1998.

[14] Alberta, "Press Release – Special Places welcomes environmental groups," 23 August 1995.

[15] Alberta, A Framework for Alberta's Special Places, no date, 20.

CHAPTER 8

[1] This excerpt is reprinted with the permission of the Stelfox family.

[2] Editorial, The Edmonton Journal, 1 May 1998. This editorial is reprinted with the permission of The Edmonton Journal.

[3] This excerpt is taken from Thomas Michael Power, "The Economics of the Oldman River Dam: A Critical Appraisal." It is reprinted with the permission of the author. See also his book Lost Landscapes and Failed Economies, (Washington: Island Press, 1996).

[4] The following references are cited by Power: Calvin L. Beale and Glenn V. Fuguitt, "Decade of Pessimistic Nonmetro Population Trends Ends on Optimistic Note," Rural Development Perspectives, 6:3, (June-September 1990) 14-18; Ken Deavers, "The Reversal of the Rural Renaissance," Entrepreneurial Economy Review, September/October 1989, 3-5; Thomas Michael Power, The Economic Pursuit of Quality, (New York: M.E. Sharpe Publishers, 1988); Gundars Rudzitis and Harley E. Johansen, "How Important Is Wilderness?: Results from a United States Survey," Environmental Management, 15:2, 1991; Gundars Rudzitis and Harley E. Johansen, "Amenities, Migration, and Nonmetropolitan Regional Development, Report to the National Science Foundation," Department of Geography, University of Idaho, Moscow, Idaho, June, 1989; Edward Ullmann, "A New Force in Regional Growth," Proceedings of the Western Area Development Conference, Stanford Research Institute, Palo Alto, California, 1955.

[5] This article originally was published in Encompass: Alberta's Magazine on the Environment, 1: 3, (July 1997). It is reprinted with the permission of the author and Encompass.

[6] This article appeared in Stan Rowe, Home Place: Essays on Ecology, (Edmonton: NeWest Press, 1990). It is reprinted with the author's permission.

[7] Wes Jackson, Altars of Unhewn Stone: Science and the Earth, (San Francisco: North Point Press, 1987).

[8] Nicholas Georgescu-Roegen, "Energy and economic myths, Part 2," The Ecologist, Vol. 5, no. 7, p. 52.

[9] This article originally was published under the title "The Elk Hunt is not the Problem," in Environment Network News, (March/April 1996). It is reprinted with the permission of the author.